Under Dark Skies

A Guide to the Constellations

Order this book online at www.trafford.com/07-1664
or email orders@trafford.com

Most Trafford titles are also available at major online book retailers.

Note for Librarians: A cataloguing record for this book is available from Library and Archives Canada at www.collectionscanada.ca/amicus/index-e.html

Printed in Victoria, BC, Canada.

ISBN: 978-1-4251-4066-3

www.trafford.com

North America & international
toll-free: 1 888 232 4444 (USA & Canada)
phone: 250 383 6864 ♦ fax: 250 383 6804 ♦ email: info@trafford.com

The United Kingdom & Europe
phone: +44 (0)1865 487 395 ♦ local rate: 0845 230 9601
facsimile: +44 (0)1865 481 507 ♦ email: info.uk@trafford.com

10 9 8 7 6 5 4 3

UNDER DARK SKIES
A GUIDE TO THE CONSTELLATIONS

BY

CHRISTOPHER LANCASTER

To Bill Dean —
May your skies by night
be clear so that the stars
shine brightly upon you!
Christopher
Lancaster
4-14-2010

For John K, who gave me the opportunity to
begin writing.

For Marcella H, whose help with this book
and with my own personal journeys
I treasure.

For my parents, Bill and Shirley, who gave
me my first eyes to the sky.

Come, gentle night. Come, loving, black-brow'd night.
Give me my Romeo; and, when he shall die,
Take him and cut him out in little stars,
And he will make the face of heaven so fine
That all the world will be in love with night,
And pay no worship to the garish sun.

--William Shakespeare
Romeo and Juliet
Act III Scene 2

Cover photograph: The constellation of Orion, the hunter. This group of bright stars stands tall during winter evenings and is easily recognized by the straight line of three stars that form his "belt".

Contents

Part One

Part Two

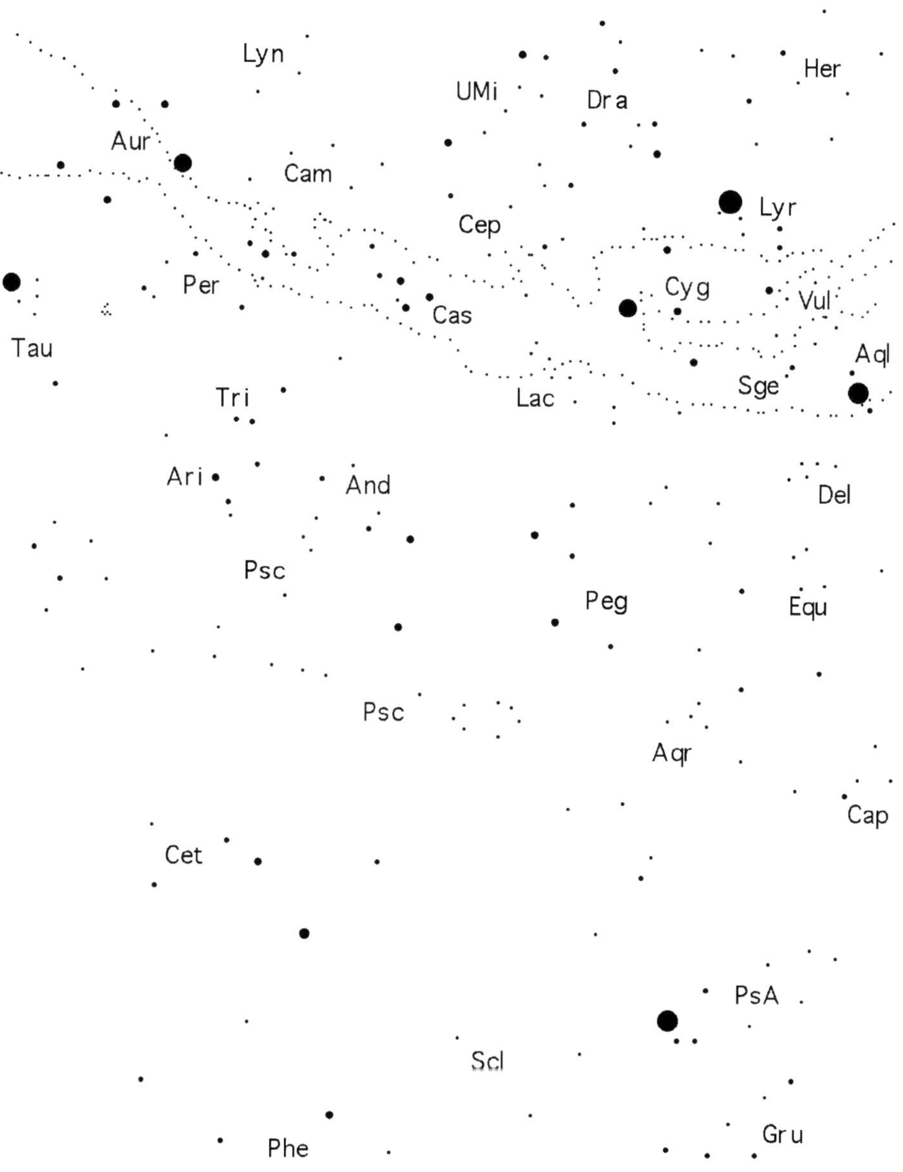

Autumn Sky Chart
10pm October 15th

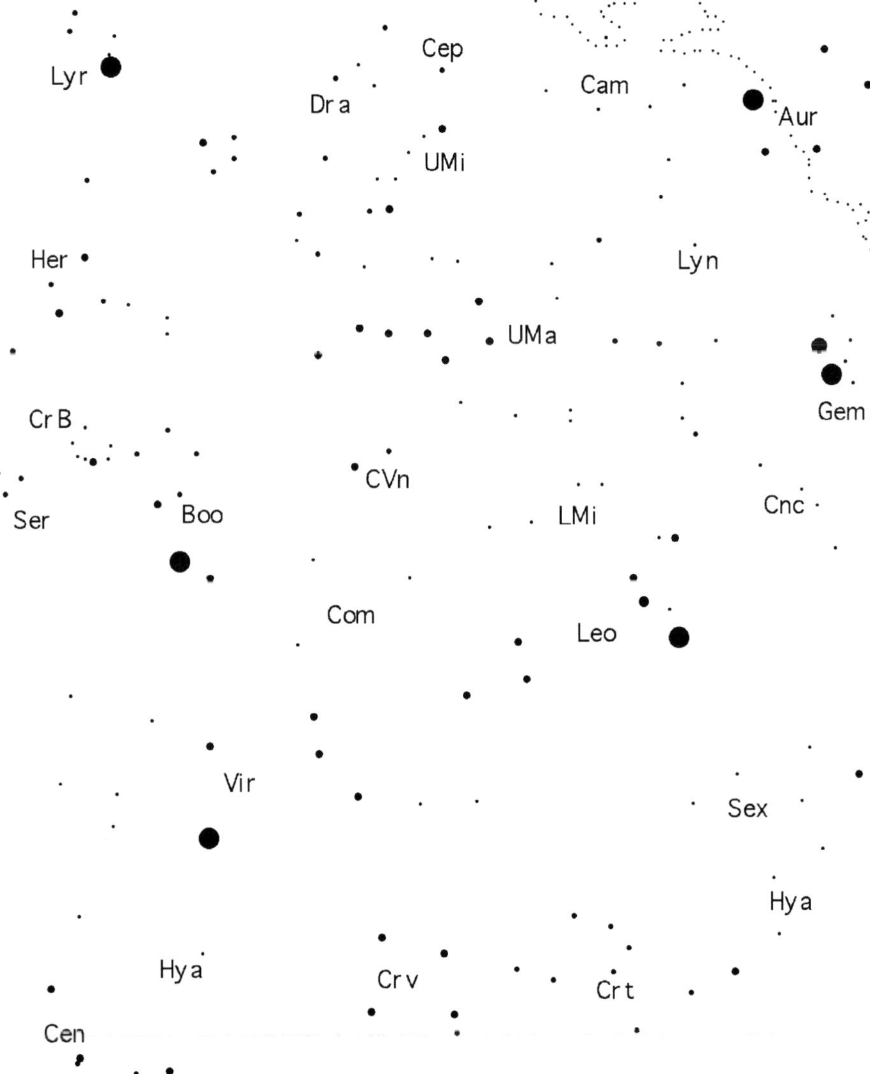

Spring Sky Chart
10pm April 15th

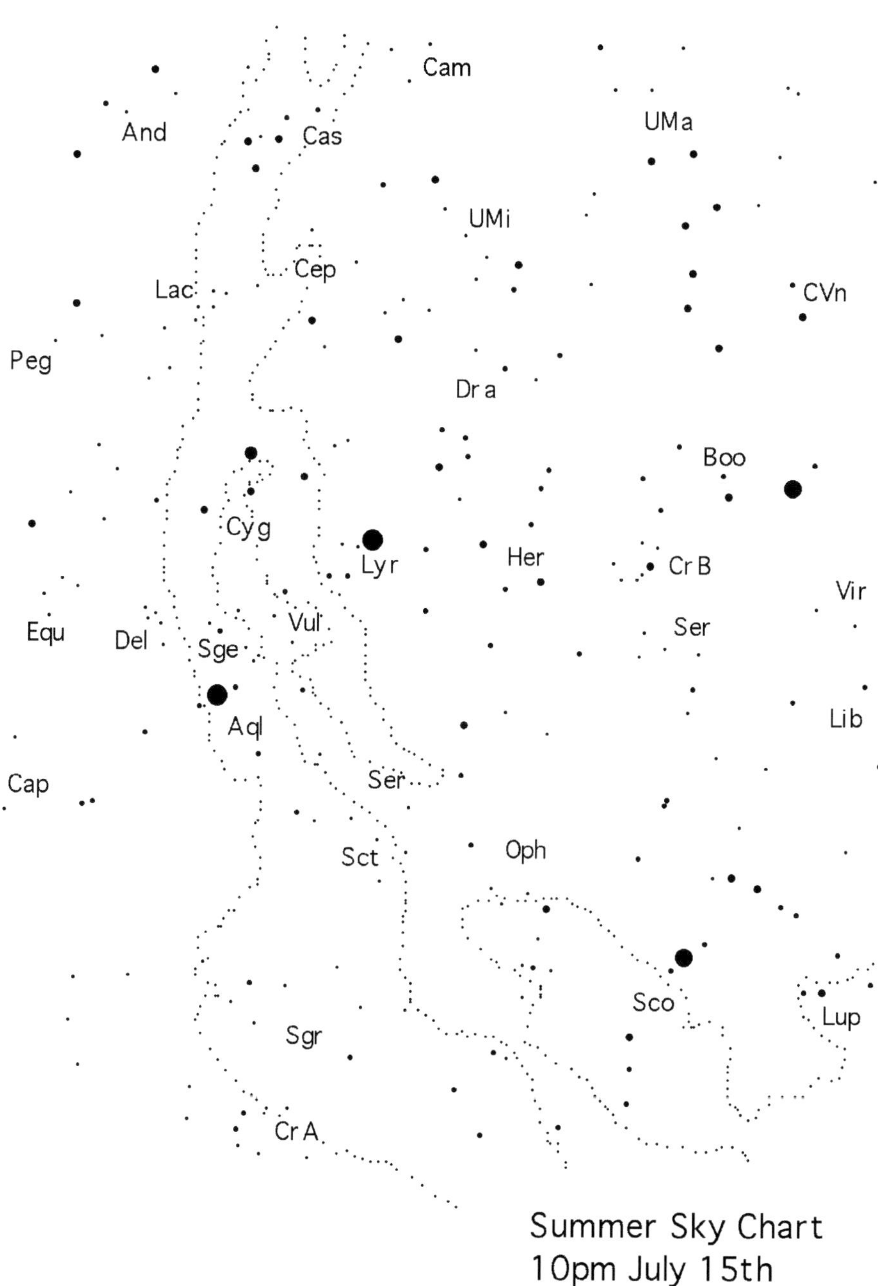

Summer Sky Chart
10pm July 15th

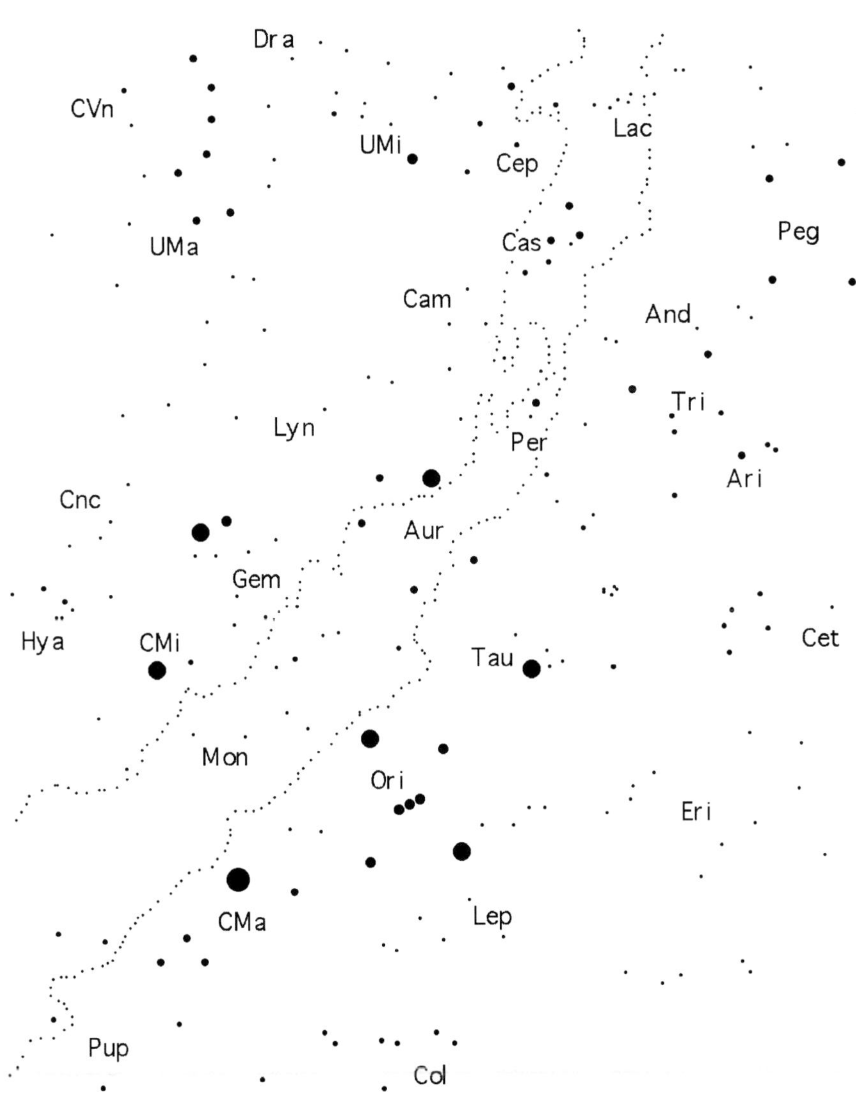

Winter Sky Chart
10pm January 15th

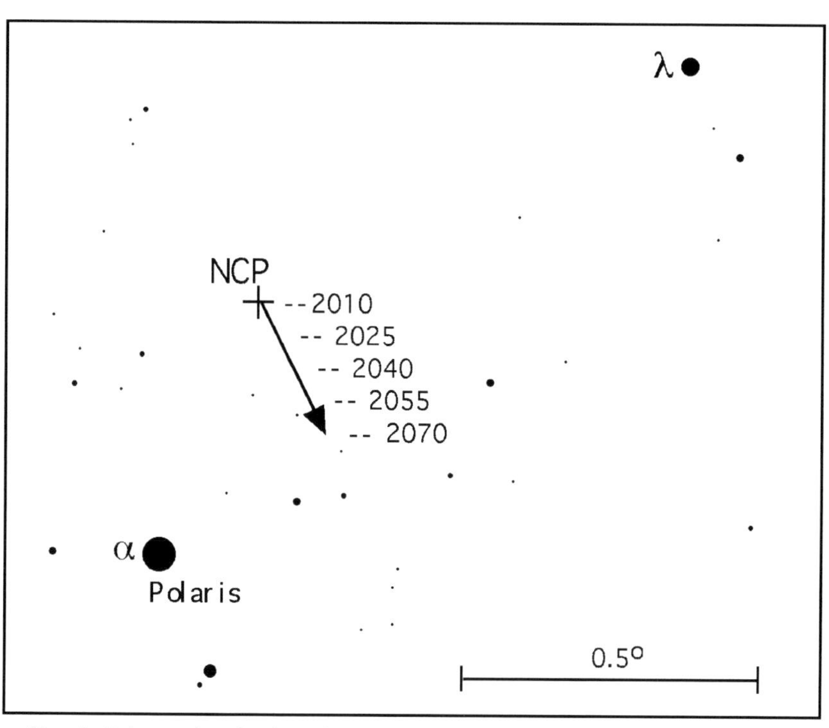

North Celestial Pole (NCP) movement during the next 60 years

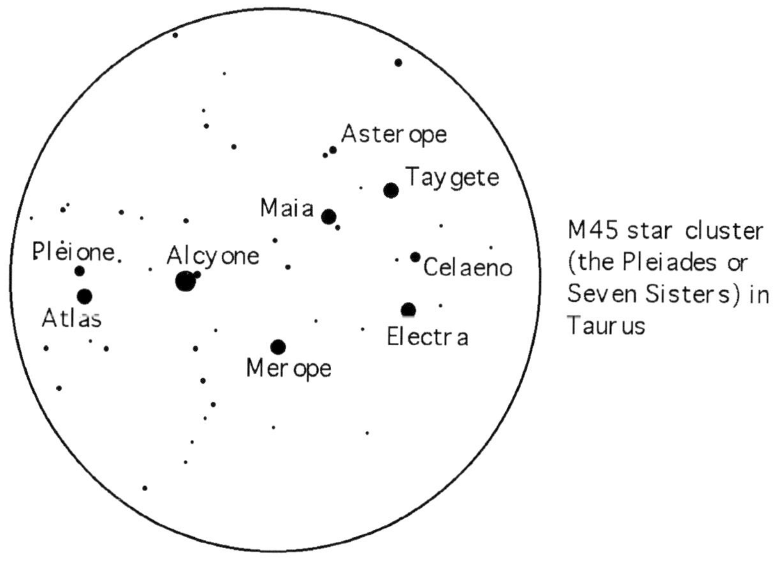

M45 star cluster (the Pleiades or Seven Sisters) in Taurus

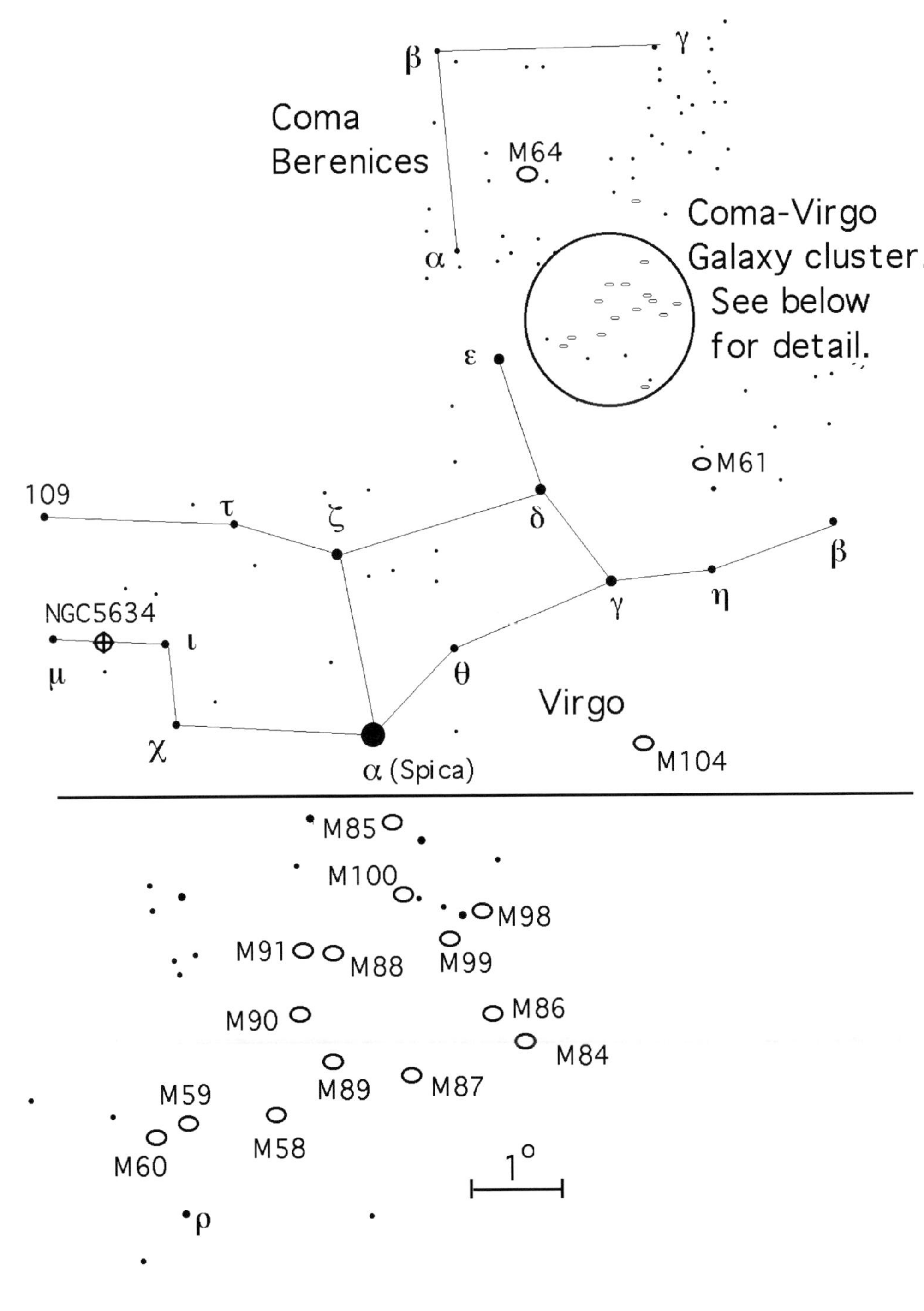
Coma
Berenices
β
γ
α
M64
Coma-Virgo
Galaxy cluster.
See below
for detail.
ε
M61
109
τ
ζ
δ
β
γ
η
NGC5634
μ
ι
θ
Virgo
χ
α (Spica)
M104
M85
M100
M98
M91
M88
M99
M90
M86
M84
M89
M87
M59
M58
M60
1°
ρ
M49

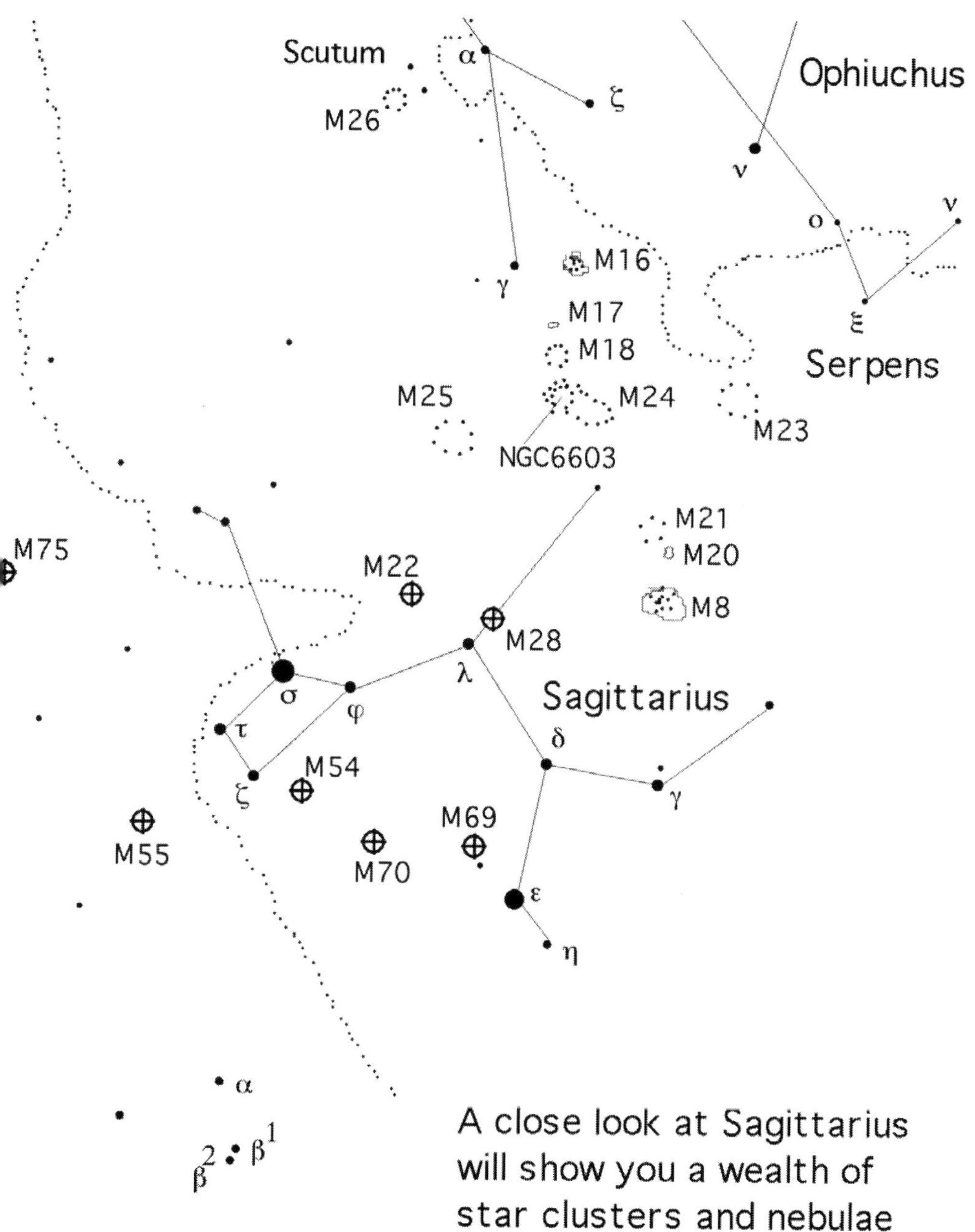

A close look at Sagittarius will show you a wealth of star clusters and nebulae

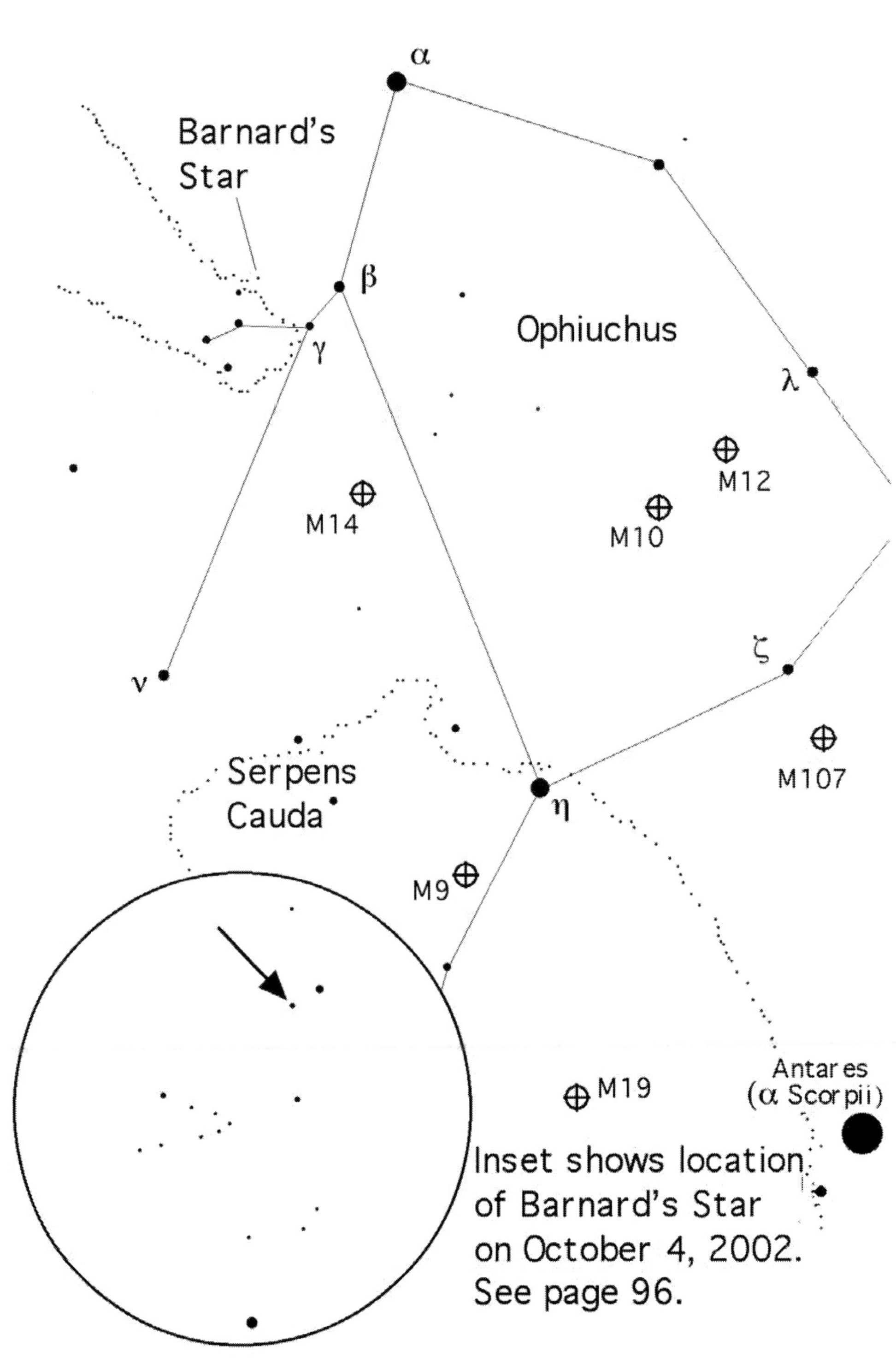

Inset shows location of Barnard's Star on October 4, 2002. See page 96.

PART ONE

BEGINNINGS

Southeast Arizona; April 14th, 2007; 8:50pm: The last vestiges of twilight had slowly faded from the horizon about an hour ago as the spot of Earth on which I was standing steadily turned away from the sun and pointed me toward distant dots and smudges of light scattered through the distant reaches of our galaxy and the neighboring expanse of intergalactic space. Earlier, I had finished setting up my telescope about a half hour before sunset and waited for dusk to fall over us to begin revealing the first bright stars in the sky as well as a passing satellite or two. It was Saturday night, close to the new moon, and along with myself, 40 or so members of the Tucson Amateur Astronomy Association had gathered here in the high desert of southern Arizona about 30 miles outside of the city for our monthly star party. Clear skies opened above us. My fellow stargazers had occupied themselves with chatter and mingling before the sun was fully down. A few had been talking about their telescopes while others were showing off the latest eyepiece they had purchased a few days ago. But now, in the darkness, it was serenely quiet as we busied ourselves with our equipment and subjects we were capturing in our eyepieces.

The warmth of the spring day had quickly yielded to evening's chill, and it would only become colder as the night continued. I slipped on a jacket as my final layer of clothing against the cold before turning my telescope to one of my favorite objects: M104, or the Sombrero Galaxy in the constellation of Virgo. I identified the constellation first by the bright star Spica, commanding attention toward the south with its blue-white light, then by looking west at the fainter stars of Gamma and Eta Virginis reaching out to the west. A bright meteor flashed across the sky, eliciting gasps from the few of us who were lucky enough to catch sight of it. I paused to watch the meteor's trail slowly fade, then continued my search. I used the LED finder attached to the tube of my telescope to point the instrument toward a triangle that I visualized with Spica, Gamma Virginis, and the seemingly featureless spot where I know the galaxy sits. I brought my eye to the eyepiece and, with a bit of panning around the dim stars in my field of view, encountered this vast island of stars 50 million light years away. The swollen center of the galaxy was sliced in two by a thin dust lane marking the galaxy's equator. Its rather small appearance in my eyepiece betrayed the fact that it was many thousands of light years in diameter and made by innumerable individual stars whose light combined into a soft glow. My earthbound reality faded for a few minutes as I absorbed the view, and I wondered if there were any life forms there looking back in my direction as well.

Wondrous thoughts go through my head each time I look at something through my telescope. It never fails to elicit fascination and awe. After almost 30 years spent in this hobby, there are things, like M104, that I've seen dozens of times and never become tired of viewing, and still hundreds of other objects which I have yet to see. I've always thought that gaining no knowledge of astronomy is to have ignored the truly vast majority of what occupies the natural firmament which we call the universe, of which the Earth, even with all of its wonder and magnificence, is only a miniscule bit of debris circling a small, anonymous star. But that is hard to comprehend with our eyes fixed to the ground. All it takes is to turn our gaze upward.

Astronomy can trace its roots perhaps hundreds of thousands of years ago when prehistoric men and women used mankind's original optical devices (their eyes) and their nascent intellects to ponder the sphere of stars that encircles the Earth. We, in modern times, can do the same. In its simplest form, stargazing requires no equipment or specialized knowledge, places no demands on its enthusiasts to travel to specific locations, and may be practiced at a moment's notice with little preparation or planning. The only essential requirement is a clear sky, which may be available to a certain degree of regularity to anyone gifted with sight. With this, the sky freely presents itself no matter where you might be on this planet's surface, since it looks the same from Kathmandu as it does from Kentucky (with perhaps some allowances given for the observers' latitude.) While you can certainly spend as much money as you desire on telescopes and their accessories, there are a few of the brightest nebulae, star clusters, and even galaxies that can be glimpsed without optical aid. This book will introduce you to what is possible in the hobby of astronomy whether you simply turn your head up to the heavens or have an elaborate observatory equipped with the largest and most technologically advanced instruments available to the amateur. If you are simply interested in learning about the constellations and where they are in the sky, then it'll be sufficient to study the star charts and the first paragraph or two of each constellation's discussion in Part 2. If you want to look closely with a telescope into each constellation to the deep sky objects each one holds, then continue reading through the discussions.

The first things you notice when looking up at the night sky are stars and perhaps the moon. The stars may number a couple dozen if viewed through the unfortunate glow of cities, or thousands of them in the darkness of uninhabited areas. As is the nature of people to "connect the dots", early sky gazers began to group together the random scattering of stars into shapes and pictures. Thus constellations came into being. Then there came fanciful stories to explain why those images appear in the sky, usually involving gods who were the only entities with the powers required to paint the black canvas of the heavens. For example, people may have seen a group of stars which reminded them of a scorpion. They reasoned that there was no coincidence in the fact that a scorpion's shape was traced in the sky, and that the gods must have drawn out the image to commemorate the story of the scorpion sent to kill the mighty hunter Orion. Or another collection of stars resembles the forequarters of a horse. Pegasus, the winged horse of mythology, was an animal which mingled with the gods, so that must be it! This fascination with explaining why the stars are where they are has been carried through the centuries by many different cultures in all areas of the Earth. This way of thinking of the sky has evolved to a more scientific study of individual objects in our universe and how and why they behave as they do. But it's fascinating to start out as a layman, or as our early ancestors did, and become familiar with the names and shapes of the dozens of constellations that adorn the night sky.

One of my earliest memories which prompted me to want to learn the constellations came from watching the movie "Airport" when it ran on television probably sometime in the middle 1970's. The scene which impressed me occurred when the airplane was in the air during their overnight flight to Rome, and the captain, played by Dean Martin, had been notified that there might be a bomb on board. The crew decided to make an emergency landing and began the maneuver to reverse course by executing a turn which was slow enough that the passengers would not notice. That would buy them time before they made a move on the bomber. However, as the captain was making an appearance through the passenger cabin, a woman stopped him because her teenage son had a question. He pointed out that he no longer saw Virgo and Leo through the window, but was now glimpsing Ursa Minor and Cassiopeia, which would indicate a drastic change in heading. At the time I didn't even know whether he was talking about constellations or the stars by themselves, but I couldn't help but think that that guy must be one of the smartest people alive to be able to recognize those things, while I would have just seen some meaningless and random spots no matter in which direction I looked. I thought it would be fantastic to know as much as that movie character did and be able to identify stars and constellations with similar ease. Little by little, I learned Orion, Cetus, Perseus, and others, as well as some individual star names like Rigel, Vega, Deneb, and Zubenelgenubi. Just like learning a language or becoming an expert in geography, learning the stars requires time. You'll want to become familiar with one constellation or area of the sky before moving on to the next one. But soon you'll be able to know the entire sky and recognize what you're looking at no matter when or where you gaze upward.

The first intent of this book is to help you accomplish that. Each constellation in Part 2 includes hints on where to find it, and the seasonal charts present an introduction to the entire night sky by showing the brightest stars and the constellations they make (identified by their abbreviations). Other charts point out a few special areas of interest. To that extent, the graphics here are limited. Detailed star maps are beyond the capabilities of these pages, and it's assumed that serious observers will already have a star atlas of some kind to use in conjunction with this book. When observing out in the field, an atlas will be necessary when looking for the specific objects that are featured on each of the constellation's pages as well as when doing your own independent explorations through the heavens.

OUR PLACE IN THE UNIVERSE

Look up on any clear night and you'll see a random scattering of stars. They come in all different colors and brightness, and are distributed unevenly across the sky. It's similar to having a handful of pebbles and then tossing them up in the air. Where they land is determined by chance. You'll see some pebbles here, others there. There will be areas where relatively few pebbles landed, and other places where, for no obvious reason, a clump of pebbles may have found their resting place. Stars represent this same kind of distribution on a three dimensional scale, although to our eyes, the dome of stars appears only two dimensional because of the immense distances involved. The sun is a star as well, floating among all those others (and taking the Earth and the other planets along for the ride) which make up a collection of hundreds of billions of stars gravitationally bound into a revolving disk shaped immensity called the Milky Way galaxy. All of the visible stars are part of the Milky Way. Their spatial relationships are difficult to conceive, but if we drastically change the scale of the cosmos to make it more manageable for our minds, we get a better idea of the universe's vast emptiness.

If our universe and everything in it were shrunk down so that the extent of our solar system within the orbit of Neptune was the size of a U.S. quarter (about 2.4 cm in diameter), the nearest star, at about 4 light years from us, would be about 108 meters away at this scale (which is equal to 354 feet, or about the distance between the back of opposite end zones of a football field). Certain telescopic objects like nebulae and star clusters (which we'll discuss later) would be 7 to 1,300 kilometers away. Our galaxy as a whole would be about 2,500 kilometers in diameter, or about the distance between Dallas, Texas, and Boston, Massachusetts. The nearest large galaxy, the Andromeda Galaxy, would sit around 55,400 kilometers away. Finally, the edge of the observable universe where ancient galaxies and quasars reside would be approximately 328,000,000 kilometers distant (which is almost three times the true Sun to Earth distance).

Most stars don't have to be very far away at all to be too distant to see with the naked eye. To put things in perspective, the Milky Way is a comparatively large galaxy. It measures close to 100,000 light years between opposite edges of its disk, and is approximately 20,000 light years thick at its center. A light year is a measure of distance, not of time. It's the distance that light covers in a year, traveling at 299,788 kilometers per second, or about 9.5 trillion (9,500,000,000,000) kilometers. Much larger distances are measured in parsecs, which is a unit of distance equal to 3.26 light years and is based on a specific parallax shift (one arc second) a star makes at that distance as the Earth moves from one point in its orbit to the opposite point as it revolves around the sun. (Parallax is the same principle that makes nearby objects seem to move position faster than more distant objects as you walk or drive down a street.) Larger distances are measured in kiloparsecs (1,000 parsecs), megaparsecs (1 million parsecs) and even gigaparsecs (1 billion parsecs). Light takes a bit more than 8 minutes to reach the Earth from the sun, and would take over 4 hours to reach Neptune (the farthest major planet from the sun). Our solar system sits about 2/3 the distance from the center of the Milky Way to the edge. Its nearest neighboring stars are what we see in the sky. Some of them are as close as 5 to 10 light years away while others are as far as 1,500 light years distant, but only the brightest stars can be seen from that far away. If we removed ourselves perhaps 50 light years from the sun, it would be just a dim, barely visible point of light in the distance.

As would be expected, we can see more stars at a "dark sky site" (a location free from the bright glow of city light pollution), but what we also see is a broad, glowing band slicing across the sky. This is the Milky Way itself. When we look at this, we are no longer looking at individual stars, but instead seeing the main body of the galaxy itself as we peer through its disk. The plane of our solar system is tilted quite dramatically with respect to the Milky Way galaxy. Therefore, the Earth's rotational axis is pointed not very far from the galactic disk, so as the Earth spins, the Milky Way's band slides across the sky to rise and set with the other stars instead of snaking across in a straight line. The stars in this galactic disk are too far away to see as individual objects, but their collective light creates a soft, glowing band wrapping most of the way across the sky. It is more prominent during summer nights, when the Earth's nighttime hemisphere is pointing toward the dense center of the galaxy, than during the winter, when we're on the other side of the sun and looking out toward the galaxy's edge.

To explain why we see these galactic features, imagine that you are a scuba diver suspended 100 feet below the ocean's surface and right in the middle of a huge school of tiny jellyfish (let's assume they are of the harmless variety), that extends up to the surface, 100 feet below you, and one mile in all other directions, which would approximate the same ratio as the Milky Way's dimensions at the sun's location. What you'd see are a few jellyfish above you, a few below, but looking to your sides, a dense cloud of the multitude of jellyfish which extends off into the distance. More jellyfish would be in your line of sight as you looked through this collection of creatures toward the edge of the school and, thus, appear more crowded, even though the density of jellyfish is the same throughout the school. We see the same thing in the sky as we look out through the top of the Milky Way in the direction of Leo and Pegasus, where stars sit in mostly inky blackness, compared to looking through the galaxy's disk adjacent to Orion and Sagittarius, where stars float against the backdrop of the Milky Way's dense clouds of stars. What this also means is that if we are interested in looking telescopically at objects which are embedded in our own galaxy, like star clusters and nebulae, look toward the path of the Milky Way. If you want to see things outside our Milky Way, which would be other galaxies, look in the parts of the sky where the Milky Way is thinnest so it doesn't block your view to these far away objects.

Even though the stars appear fixed in the sky, they are all moving. A complex interaction of gravity and stellar motion holds the Milky Way galaxy together. The sun is moving around the center of the galaxy at about 200 kilometers per second, but the gravity that the galaxy exerts on us keeps us from flying away into intergalactic space. All the other stars around us are similarly circling through the galaxy, but since some are slower, and some move more quickly (and in slightly different directions), they all have a motion relative to the sun. The independent movement of stars across the sky is called proper motion, and is a manifestation of the stars' intrinsic motion combined with its direction through space relative to the sun. Such motions toward or away from the sun range from 1 or 2 kilometers per second to over 100 kilometers per second for the fastest moving nearby star. For example, the sun is moving about 20 kilometers per second in the direction of Vega, the bright star in the constellation Lyra. This sounds like a huge speed, but the vast distances between the stars make their relative positions change very little and seem fixed during a human lifetime. They move across only very small fractions of the sky even over hundreds of years. Given hundreds of thousands or millions of years, however, all the stars in the sky will have moved into different positions, making the constellations appear very different from how they look today.

To add to the complexity of the sky, where the Earth is in its orbit determines what constellations we see at a certain time of night. When the Earth is on one side of the sun during summer, for example, its nighttime hemisphere is pointing in a direction that is opposite to when the Earth is on the other side of the sun during the winter. Say you are standing in a field of central Nebraska at midnight and looking at a sky filled with stars. If it's summer, you'd be looking at Sagittarius, Lyra, Cygnus, and other "summer" constellations. Cut to 6 months later when the Earth is on the opposite side of the sun (imagine if you were able to grab the Earth in place and relocate it). If the Earth's hemispheres are pointing in the same direction, it would now be noon. The same constellations that we saw earlier would then be hidden behind the sun, and if you waited 12 hours, it would again be midnight and you'd be looking in the opposite direction toward "winter" constellations such as Orion, Canis Major, and Auriga. Incidentally, this illustrates the difference between Earth's sidereal rotation (with respect to the fixed stars) of 23 hours, 56 minutes and 4 seconds compared to its solar rotation of 24 hours. To compensate for moving around the sun during the course of a day, it takes the Earth an additional 3 minutes and 56 seconds to turn a little bit more in order to put the sun in the same place in the sky as the previous day, while the celestial sphere has shifted slightly during that same interval. Over the course of six months, this difference adds up to 12 hours.

NAMING THE STARS AND CONSTELLATIONS

As you become familiar with the shape of the constellations, you'll see that some of them look very much like the object or animal they are supposed to represent, while others require a good imagination for you to come close to visualizing. Many of the constellations recognized today were first envisioned in ancient times. Some constellations inspired different names and stories among the varied cultures spread across the world, as would be expected, but others brought about remarkably similar tales. These were of things and stories that were familiar at the time, such as animals and mythical men and creatures. We find these old constellations in the northern hemisphere where most astronomical records originated. When Europeans began traveling to the southern hemisphere beginning in the 16th century, they saw an entirely new set of stars and constellations, and these were given names mostly of things which played a prominent role in the developing civilization of that time (such as ships, microscopes, furnaces and air pumps). Up until fairly recently, a group of stars could be given a name of anyone's choosing, and different star maps showed different constellations depending on who created the map. But in 1930, astronomers across the world decided to standardize the sky by officially establishing permanent names and boundaries for the constellations that exist today. What we have now are 88 constellations with Latin or mythological names scattered across the sky in a sort of patchwork, and each has a well-defined area that is separated from the others with linear boundaries. All areas of the sky are claimed by one constellation or another, so every star and every astronomical object sharing the sky belongs to one specific constellation.

It's been established that the constellations have names, and in addition, some of the individual stars that make up these constellations have separate names of their own. The Greco-Roman civilization compiled a vast amount of information on the heavens, and upon its fall, that knowledge landed in the hands of the Arabs who recorded that information in their language. When the Europeans in turn collected this information during the Renaissance, they kept most of the Arabic star names for their star atlases, and these names have survived to the present. Thus, you'll see names such as Mintaka, Hamal, Menkalinan, Algenib, and a host of others. The translations of which normally have something to do with the stars' constellation or the part of the constellation the stars form. A smaller number of star names are Greek, Latin, or Hebrew in origin.

Another way of naming stars appeared in 1603 in Johann Bayer's star atlas, Uranometria. He decided to give the stars that form a constellation Greek letter designations, generally going in order of each star's brightness or prominence in the constellation as he went through the alphabet, followed by the possessive form of the constellation's name. Therefore, one star is named after the first letter of the Greek alphabet, alpha, another star after the second letter, beta, and so on. That allowed up to twenty-four stars in any individual constellation to be named in this way. So, for example, the "third" star of the constellation Pisces is Gamma Piscium (or γ Piscium). The "eleventh" star of Orion is Lambda Orionis (λ Orionis). The letters that comprise the Greek alphabet are as follows:

α Alpha	ν Nu
β Beta	ξ Xi
γ Gamma	ο Omicron
δ Delta	π Pi
ε Epsilon	ρ Rho
ζ Zeta	σ Sigma
η Eta	τ Tau
θ Theta	υ Upsilon
ι Iota	φ Phi
κ Kappa	χ Chi
λ Lambda	ψ Psi
μ Mu	ω Omega

Since there are many bright stars in the larger constellations and the list of Greek letters can be quickly exhausted, the British royal astronomer John Flamsteed used numbers several decades later to indicate stars in a constellation, using higher numbers the farther east in a constellation you look, or in order of when stars cross a fixed north to south line in the sky as the Earth rotates. Using this system, we'll see such stars as 71 Cancri (in Cancer) and 46 Ceti (in Cetus), for example. All of these naming conventions will be seen on a good star chart. Other methods of naming stars come from more comprehensive catalog listings and include the Henry Draper (HD) catalog, the Harvard Revised (HR) catalog, the Smithsonian Astrophysical Observatory (SAO) catalog, and several others, which all use numbers to identify stars. Moreover, variable stars have their own naming system using either single or double letters, such as V Aquilae (in Aquila), R Leonis (in Leo), and SU Tauri (in Taurus). All of these naming systems can come together to give some of the individual bright stars many different designations. If we pick a star at random, let's say the star marking the middle of the neck of Leo, the lion, we'll see that this star is identified by the names of Adhafera, Zeta (ζ) Leonis, 36 Leonis, HD 89025, SAO 81265, HR 4031, IRAS 10139+2340, as well as several other, more obscure listings.

ORGANIZING THE SKY

In geometry, a full circle measures 360 degrees, and so does a full circle traced around the sky that divides it into two equal halves. Each degree is divided into 60 minutes of arc (or arc minutes, often indicated as the symbol '), and each minute of arc is divided into 60 seconds of arc (or arc seconds, indicated as the symbol "). This comes in handy when measuring how far apart two objects are in the sky, such as the separation of double stars or how far away a deep sky object is from an easily located bright star.

When indicating an object's exact location in the sky, the simplest way is using altitude and azimuth. An object's altitude is its height above the horizon in degrees from 0 to 90, with 0 being on the horizon, and 90 at the zenith. Azimuth is expressed also in degrees, but this is the horizontal axis with 0 being true north and increasing up to 359 as you turn in a circle clockwise. So, for example, if you face north and rotate in azimuth 59 degrees 47' to the right (to point east northeast), and look up 63 degrees 10' in altitude from the horizon, you'd be looking at a precise point in the sky from your vantage point. This is a less formal way of indicating an object's position, since both the altitude and azimuth of that object for two observers at the same moment will be different depending on each observers' location on Earth. It will also change as time goes by and the Earth rotates below it.

A method independent of such variables borrows the same concept as latitude and longitude on the Earth's surface. The imaginary lines that are drawn on the sky that correspond to Earth's latitude are called Declination, expressed in degrees, minutes, and seconds, and indicate how far north or south an object is from the celestial equator. The lines analogous to Earth's longitude are called Right Ascension, expressed in hours, minutes, and seconds, and indicate how far an object is east or west of a certain point. These are fixed lines in the sky. The celestial equator is 0 degrees Declination (abbreviated Dec), the north celestial pole near the star Polaris in Ursa Minor is +90 degrees, and the south celestial pole, -90 degrees. Right Ascension, (abbreviated RA), is measured in hours, minutes, and seconds, and goes from 0 hours to 23 hours, with the 0 line running through the point where the sun is located during the vernal (spring) equinox. Both the degrees of Declination and the hours of Right Ascension are divided into 60 minutes, and each of those minutes are divided into 60 seconds, although you should take care not to make an identical comparison between the two. The minutes and seconds of Declination are much smaller than those of Right Ascension, since the former has 360 units around the celestial sphere while the latter has only 24 units.

To see an example of the coordinates of an actual object, let's use one of my favorite globular clusters: M15, which is near the nose of the mythical flying horse envisioned in the constellation of Pegasus. Its celestial coordinates are RA 21h 30m Dec +12° 10'. This means that it is at 21 hours 30 minutes in Right Ascension, and 12 degrees 10 minutes north of the celestial equator. One way to think of Right Ascension with this example is this: if you pointed a telescope at 0 hr RA and at the correct declination, 21 hours and 30 minutes later, M15 would be centered in your eyepiece. To determine how far an object can be in the southern sky and still be above your horizon, simply subtract 90 from your latitude. This will give you the southern declination that will be even with the horizon at true south. For example, being in Chicago (with a latitude of 42° N) will enable you to view objects down to -48° declination. From Miami (with a latitude of 26° N) you'd see down to -64° declination. Objects with declinations farther south of these values will always be below the horizon for residents of these cities. A point to remember when dealing with this type of celestial mapping method is that an object's RA and Dec will change as the Earth goes through its precession cycle, which is the roughly 26,000 year wobble that our planet's rotational axis undergoes as a result of the gravitational tugging of the sun and moon. An object's precise RA and Dec will be slightly different now than it was several years ago. While this is a slow change which is constantly occurring, objects' locations will be updated every 50 years in astronomical publications. Older publications will show coordinates which were accurate in 1950 (Epoch 1950) and were adequate for use in locating an object during a span of a couple of decades before and after that year, while publications published in the 1980's, for example, began to use RA and Dec numbers based on Epoch 2000 positions. Once we approach the middle of this century, Epoch 2050 coordinates will come into use.

In addition, there are two major imaginary lines in the sky which are prominent. One is the meridian. This is the line running from true north to true south and through the zenith, dividing the sky in two equal halves. AM and PM, the suffixes added to the time of day, are derived from this line when the sun is traveling toward it in the morning (ante-meridian) and after it has passed it in the afternoon (post-meridian). The other major line is the ecliptic. This marks the plane of our solar system and is the path along which the sun, major planets and minor planets move. The constellations of the zodiac are lined up along this ecliptic line.

WHAT DO THE CONSTELLATIONS HOLD?

Once you have learned the constellations, you might want to know what objects might be interesting to observe up close through the optics of a telescope. Some things you can see with your unaided eye, but to reveal most of the fascinating things that float out there in space in any detail require either binoculars or a telescope of some kind. To some people's surprise, a good pair of binoculars, which are actually two low power telescopes used in unison, can capture excellent images of the larger and brighter astronomical objects. Even the moons of Jupiter can be glimpsed through standard power binoculars when they are held steady. So, what is up there to behold? Of course the moon and planets are prominent on any night that they are above the horizon. The planets are constantly moving, so they may require some initial effort to track down and identify. Filling most of the sky are the fixed objects: stars (which come in all kinds of varieties and characteristics) and what are called deep sky objects. These include nebulae, star clusters, and galaxies, each with their own different types. Let's begin to explain these different types of objects.

STARS

These are by far the most abundant objects that are seen in the night sky with either just your eyes or with standard backyard equipment. They come in all varieties, including double or multiple stars (a close pairing or group of stars in orbit around each other), and variable stars (those that change brightness over time), both of which are discussed in more detail below. The most obvious characteristic of stars seen at first glance is that they all have different brightness. The earliest astronomers categorized the stars that they saw by how bright they are. The brightest stars were said to be of the first magnitude, the next brighter stars were of the second magnitude, and so on. This system is still used today, but adjusted to a more scientific level. The brightness of stars is measured on a logarithmic scale, with the term magnitude still used. The brightest stars are of magnitude 0 and below, with Sirius, the brightest star in the sky, shining with a magnitude of -1.5. Stars with diminishing brightness are magnitude 1, 2, 3, 4, etc., and fractions in between. As you can see, the higher the number, the dimmer the star. From one magnitude value to the next, the difference in brightness is 2.512 times. A five magnitude difference equates to a difference in brightness of 100 times (2.512 to the power of 5). The dimmest stars you can see with the unaided eye from a city location are between 3 and 4. From a very dark site, the dimmest is about 6. A backyard telescope, in addition, will allow you to spot stars down to 12th magnitude and dimmer.

The magnitude scale explained above is called apparent magnitude, because it tells us how apparent a star's light appears to us from our observation point on Earth. However, it doesn't tell us how much total light the star emits since a star that is farther away will look dimmer to us than an identical star that is closer. For this, we use another term: absolute magnitude. This value is assigned for each star by determining how bright it would appear to us from a standard distance, which is established at 32.6 light years away (or 10 parsecs). Our star, the sun, shining with an apparent magnitude of -26.7, is only so bright because it is just a few light minutes away. However, it has an absolute magnitude of 4.8, which is how bright it would appear if it were 32.6 light years away. There are stars that are close by and therefore appear bright in the sky, such as α Centauri, but emit only slightly more light than the sun; and there are others that are fairly dim to our eyes, but since they are very far away, are extremely bright in their absolute magnitude. One example of this is Deneb, the alpha star in the constellation Cygnus. It's still a fairly bright star to our eyes shining at an apparent magnitude of 1.25, but its absolute magnitude is -7.2. It shines about at least 60,000 times brighter than the sun. When we talk about non-stellar (or diffuse) objects such as star clusters, nebulae, and galaxies, the assigned magnitude is determined by how bright it would appear if all of its light were compressed to a single point.

Another thing you might notice about stars, if you look closely, is that they exhibit different color. We've talked about how stars differ in brightness, but they also differ in size and temperature. A star's temperature is what determines the color, or spectral type, that it exhibits. Just like putting a piece of metal in a furnace, it glows red as it heats up, and then changes to orange and eventually white or bluish if you are able to heat it up enough. Stars show this same phenomenon. Red and orange stars are the coolest, yellow-white stars like the sun are in the middle, and the hottest stars glow white and blue. A normal star's temperature typically corresponds to its mass. Generally, the more massive the star, the hotter it is because the gravity created by this mass squeezes the star's core to higher pressures and temperatures. A star's mass is measured in solar masses. This unit is established according to the mass of our own sun, which is 1 solar mass. Stars range from fractions of a solar mass for the coolest stars to over 100 solar masses for the heaviest and hottest stars. Interestingly enough, stars of higher mass have shorter life spans. Even though there is more hydrogen fuel to make it shine, the higher mass creates a more energetic environment in the star's core, making it burn through its quantity of hydrogen faster. While stars like our sun will shine for up to 10 billion years, very high mass stars will exhaust their quantity of hydrogen in only a few hundred million years or less. Stars' colors are more apparent through a telescope, but the most colorful stars like red Betelgeuse (in Orion) and Antares (Scorpius), orange Arcturus (Boötes), and bluish-white Rigel (Orion) and Spica (Virgo) show their colors well to the naked eye. Stars are categorized according to their spectral type by the following letters:

Type	Color	Surface temperature in degrees Kelvin*
O	Blue	28,000-45,000
B	Blue	10,000-28,000
A	Blue-White	7,500-10,000
F	White	6,000-7,500
G	Yellow-White	5,000-6,000
K	Orange	3,500-5,000
M	Orange-Red	2,000-3,500
N	Red	<2,000

*the range of a Kelvin degree is equal to a Celsius degree but the Kelvin temperature scale begins at absolute zero, or -273 degrees Celsius.

The classic way to remember the order of the letters on this spectral type list is by using the phrase "Oh, Be A Fine Girl (or Guy), Kiss Me Now." Each letter designation also has different subtypes from 0 to 9, with higher numbers indicating cooler temperatures within the letter type. Rigel, the hot blue-white star in the constellation Orion for example, is of spectral type B8. Antares, the reddish orange star in Scorpius, has a spectral type of M1. An experienced observer can roughly guess a star's spectral type if its color is noticeable. Furthermore, stars are given luminosity types using the Roman numerals I through VII according to the following list:

Supergiants	Ia, Iab, Ib
Luminous giants	II
Giants	III
Subgiant	IV
Main sequence	V
White dwarfs	VI, VII

Therefore, the sun is categorized in its entirety as G2V, meaning its spectral type is G2, having a surface temperature of about 5,800 degrees Kelvin and is a main sequence (or stable, middle aged) star.

Finally, stars have different sizes, but that is of little interest to the casual observer aside from a curiosity standpoint. Since stars are so far away, even the highest magnification possible with the current optical science available to the amateur astronomer cannot truly resolve the size of a star. They will all look like points of light. However, it is interesting to know that the sun is considered a dwarf star, even though its diameter is over 1.2 million km. White dwarf stars can be as small as the Earth, while the biggest stars, like Betelgeuse in Orion, are close to the size of the orbit of Mars, or 456,000,000 km in diameter.

DOUBLE AND MULTIPLE STARS

Many stars are either members of a pair (called double or binary stars), or a group (called multiple stars). These are bound together by gravity in stable orbits. Through a telescope using various magnifications, you'll be able to see each member of many such doubles and groups. When describing a double star, the brighter star is called the primary, and the dimmer one, the secondary. The two stars of a binary pair can either be almost identical in brightness and color, or vastly different. The position angle (PA) of a double star describes the location of the secondary star with respect to the primary. This angle is measured from 0 to 359 degrees moving eastward in a circle from north, which is the same as azimuth, but in this case would be in a counterclockwise direction when looking up. The separation of a double star gives a measurement of how far apart they appear in our Earth bound sky, expressed in arc minutes or arc seconds. While most of these stars are true doubles and multiples, there are a few that are visual doubles. These are stars that appear to be pairs, but only because they lie in the same line of sight and are separated by vast distances with no physical connection at all. The closest separation that can be comfortably resolved by a medium sized telescope (such as 8 inches) is about 2 arc seconds.

This is approximately the apparent diameter of a U.S. quarter when viewed from 2.4 kilometers away.

VARIABLE STARS

Unlike the sun, which shines with a steady light, there are many types of variable stars. These change brightness over different periods of time. They are summarized below.

Eruptive variables: stars that change brightness due to explosions, such as novae and supernovae.

Pulsating variables: the change in brightness of these stars is a result of the star periodically expanding and contracting, much like a balloon if it were repeatedly inflated and deflated. As the star expands, its apparent magnitude becomes brighter.

Long period variables (LPV's): old, cool stars that change in brightness over a wide magnitude range over a span of 3 months to two years or more.

Cepheid variables: a class of variables named after delta Cephei, the first star of this type to be discovered. They change in brightness up to two magnitudes during periods ranging from 1 day to 10 weeks. Since their periods are directly related to their absolute magnitude, their distances can be accurately measured.

RR Lyrae variables: hot, short period variables that complete a cycle in 1 to 25 hours. These are named after the first star to be discovered of this type of variable in the constellation Lyra.

Irregular variables: stars of various kinds which follow no predictable pattern as they change in brightness.

Eclipsing binaries: while this type is included among the different kinds of variables, these are not true variable stars since their variations in brightness only come from one star eclipsing another star in a binary system. They only appear to vary in brightness because their orbital plane is oriented edge-on to our line of sight, and as one star passes in front of the other, some or all of the background star's light is blocked, resulting in the drop of the total light output of the system.

STAR CLUSTERS

"Star cluster" is the obvious name given to a collection of stars grouped close together. They come in two varieties: open (also called galactic) clusters, and what are called globular clusters. Open clusters are within our own galaxy and contain anywhere from a mere couple dozen stars to many thousands. They mainly lie in the swath of the Milky Way during the summer and winter evening skies. Some open clusters are very young and can still be swaddled in the nebulous gasses from which they commonly formed. Due to their comparatively small size (ranging from perhaps as little as a dozen stars up to a few hundred), open clusters don't have adequate gravity to hold them together, so over time, their stars often disperse to travel independently through space. Globular clusters, on the other hand, are old, giant clusters, almost as old as the Milky Way itself, and found in a large sphere surrounding our galaxy. They can, therefore, be seen in almost any direction. While open clusters can be irregular in shape, globular clusters are spherical or subtly oval in shape and can contain many millions of stars.

NEBULAE

A nebula (Latin for "cloud") is a collection of hydrogen or other elemental or molecular gasses and dust contained within a galaxy. There are five types of nebulae as described below:

Emission nebulae shine with their own light as one or more energetic stars within the nebula impart radiation to the nebula's hydrogen and make it fluoresce, similar to what happens in a neon light tube when high voltage electricity flows through it. Emission nebulae commonly exhibit a reddish or pinkish glow when captured on film or CCD.

Reflection nebulae do simply what their name implies. They are not visible by their own light but instead reflect light from nearby stars, making them show hues of blue and gray.

Dark (or absorption) nebulae block light coming from background stars or other nebulae. They manifest themselves as featureless voids, almost like a hole punched through the star field.

Planetary nebulae have the name they do not because they have anything to do with planets. Instead, they are typically small and spherical, and therefore resemble planets when viewed through a telescope at high power. They form when a star reaches the end of its life and expels its outer atmosphere in an expanding sphere of glowing gas which can become several light years in diameter before finally dispersing into space. Our own sun will become one of these nebulae once its nuclear fuel is exhausted billions of years in the future.

Supernova remnants are our last type of nebula. They are created after a massive star explodes. The aftermath is an irregular cloud of debris expanding into space.

GALAXIES

Galaxies are the largest conglomerates of stars, dust, gasses, and everything else that is conceivable in the universe. They range in size from a few thousand light years to over 100,000 light years across, and all of that material organizes itself into a a variety of different shapes. The three major types of galaxies are elliptical, spiral, and irregular.

Elliptical galaxies are spherical to oval in shape and are categorized from E0 (spherical) to E7 (markedly flattened). Spiral galaxies have the classic shape that most people, probably, consider a galaxy to have. They are disk-shaped with a central bulge and arms wrapping around the bulge. They are designated S0, Sa, Sb, and Sc. S0 galaxies have indistinct arms, as if they are wrapped so tightly that they blend in with each other. S0 galaxies are also known as lenticular (or lens shaped) galaxies. Their appearance is somewhere in between true ellipticals and spirals. Sc spirals have very loosely wrapped arms, and Sa and Sb spirals are somewhere in between. (Barred spiral galaxies, having a conspicuous lane of stars traversing their centers, are similarly categorized SBa, SBb, and SBc.) The Milky Way is a spiral galaxy, thought to be a SBb type. Finally, irregular galaxies are small galaxies that have no organized shape at all. The Large Megellanic Cloud and Small Megellanic Cloud, which are satellite galaxies of the Milky Way that are visible from the southern hemisphere, are examples of irregular galaxies.

DEEP SKY OBJECT CATALOGS

As you read about the constellations later in this book, you'll notice that their featured objects are identified by what look like code names made of letters and numbers. The letters indicate the catalog from which the designation is taken, and the numbers are the catalog numbers for the particular object. There are three major catalogs in use: Messier, NGC, and IC.

The Messier catalog, compiled by the French astronomer Charles Messier in the 18th century, contains 110 objects that can be easily viewed through amateur telescopes, such as our example of M15 above (M for Messier, and this as his 15th catalog entry). These are the most popular objects among amateur astronomers since they are also the most spectacular and easy to find objects in the sky. It's interesting to know that Messier was a comet hunter, and the reason for making his list of these deep sky objects was so that he could disregard them on future nights if they again crossed his eyepiece as he continued his search for comets.

After Messier's time, as telescopes became larger and their optics more advanced, more objects were seen and discovered. The "General Catalog of Nebulae and Star Clusters" (at the time, galaxies were not recognized for what they are and were thought to be nebulae) was compiled by the British astronomer Sir William Herschel in the 19th century. In this listing, Herschel recorded hundreds more of these objects. Subsequently, J.L.E. Dreyer, a Danish immigrant to Ireland, was recording many more objects of his own over the next 20 years after

Herschel. From the urgings of the Royal Astronomical Society, he arranged Herschel's and his own list of astronomical objects into the "New General Catalog of Nebulae and Star Clusters" (NGC), which lists more than 5,000 of these deep sky wonders. Additional objects were added to the NGC list around the turn of the 20th century as the Index Catalog (IC).

A detailed star chart will show objects labeled with such names as M67, NGC4565, and IC2062. All of these cataloged objects comprise all the different kinds of deep sky objects introduced earlier: star clusters, nebulae, and galaxies. A fairly recent list made exclusively of galaxies was published in 1989 and is called the Principle Galaxy Catalog (PGC).

OBSERVING HINTS

ENTERING THE NIGHT

After the sun sets, we enter a period of twilight before the sky is fully dark. This lasts about 80 minutes for southern temperate and subtropical latitudes and about two hours for high northern temperate latitudes. This difference in the length of twilight is made by the sharper angle the sun follows toward the horizon for observers in high latitudes compared to those farther south. Therefore, as the sun sets, even though its speed across the sky is always the same for all observers, the distance below the horizon increases more slowly for residents of Canada and northern Europe compared to those in Mexico, Central America, and other locations that are closer to the Earth's equator. This twilight is divided into three periods. Civil Twilight is first, lasts about 30 minutes, and begins when the top of the sun's disk disappears beneath the horizon. During this time it is still bright enough to conduct normal outdoor activities without artificial light sources. The end of Civil Twilight occurs when the sun has sunk 6 degrees below the horizon, and then we enter the period of Nautical Twilight, which lasts for the next 30 to 40 minutes. By then, the brightest objects in the sky (including Venus, Jupiter, Saturn, and some of the brightest stars) have become visible. As we sink deeper into Nautical Twilight, the dimmer, outlining stars of many of the constellations begin to emerge while the horizon is still distinguishable. The appearance of stars during this period allows ocean vessels to navigate by them, which is why the term originated. At the beginning of the 30 to 50 minute period of the last phase of twilight, called Astronomical Twilight, the sun is 12 degrees below the horizon, and most of the remaining background stars have come into view. The last hint of sunlight then fades from the horizon by the time the sun is 18 degrees below it. For most purposes, the sky can be considered completely dark for most of the time spent during Astronomical Twilight, but there is still enough light scattered from the horizon to interfere with the observation of the dimmest celestial objects. As the next day begins, this process unfolds in reverse as the sun prepares to rise again.

When planning to go out for a night of observing, allow enough time for your eyes to adjust to the darkness. If you are out before sunset, then they will do this adjustment as the sky darkens. But if you go straight from a brightly lit indoor environment to your backyard at night, then it will take your eyes approximately 30 minutes to acquire their full night vision and enable the best observation of dim celestial objects. This is because one type of light sensitive cell that comprise your retina produces a pigment called rhodopsin (also known as visual purple). This substance makes them more sensitive to light. However, it breaks down fairly quickly in the presence of bright, white light. In low light or darkness, rhodopsin builds up during this 30 minute interval to a level sufficient for good night vision. Therefore, it's important to stay in a dark environment during your observing session unless you have another 30 minutes to devote to fully regain your dark adapted eyesight.

Once this is attained, even pairing the best telescope with the most experienced observer is useless in certain conditions. Just like rain isn't conducive to a good baseball game, certain weather conditions will determine whether you'll have a good night of observing. Of course, a cloudy night is an obvious impediment to observing, and avoiding the moon will prevent it from drowning out the dim objects you might want to view. Moreover, wind will jiggle your telescope so badly that your images will shake too much for good viewing. But two other conditions make for either good or bad observing. The first is "sky transparency". This is a measure of how much light will be able to pass through the Earth's atmosphere to make it to your eyes. Factors that determine this are presence or absence of haze, dust, fog, or high clouds. If most of the dimmer stars are not visible when you look up in the sky, then the transparency is low and your quality of observing dim objects will also be low. Conversely, if you are able to see down to 5th or 6th magnitude at a dark sky site, then the transparency is good.

The other condition is called "seeing." We are all familiar with the fact that stars twinkle. This is because their light is passing through different layers of the Earth's atmosphere where turbulence unevenly refracts the previously uniform beam of light coming from a star. If there is a large degree of instability and turbulence through the Earth's atmosphere, then stars will have a great deal of twinkle, or scintillation. Likewise, a star's appearance through a telescope will also exhibit this scintillation, becoming worse with high magnification. Therefore, the best night of observing will come in a sky that allows the light of dim stars to pass through to the bottom of Earth's atmosphere, and makes the stars twinkle to the least degree. Of course there are a couple of things that you can do to help yourself to steady skies. Avoid looking across hot surfaces like parking lots and rooftops, or through open windows when there is a large temperature difference between inside and outside. View objects when they are the highest in the sky, as well, so you will be looking through the least amount of atmosphere. Viewing something that is just above the horizon will be like trying to see something well through the heat waves of a campfire. It'll shimmer too much to see clearly. Finally, the later it is at night, the more time the air above you has had to settle down for good seeing.

If you are looking at very dim objects, then there are a couple of tricks that you can do to bring the objects into better view. One is to not look directly at the object in your eyepiece, but to look to one side. This is called averted vision. (I've heard the joke from a few astronomers who call it averted imagination if the object is still not visible!). To better understand how this works, let's return to the cells that make up your eye's retina. There are two different types, called cones and rods (the latter kind are the type mentioned earlier that produce rhodopsin). The cones detect color and discern a higher resolution image that your optic nerve sends to your brain. Rods see in black and white and make a lower resolution image, but are more sensitive to light. That's why in very low light conditions (like deep in twilight), only your rods are stimulated to send an image to your brain. The result is that you can still see things, but it's hard to make out details. An example of this is being able to tell that you have an open book in front of you, but you can't read the words. Your retina is designed with a higher concentration of cones at the point where the image collected by your eye is centered, called the fovea. That enables you to see more details when you look directly at something. But since it's the cones that are recording the center of the image you are seeing, the view can be "underexposed" or even invisible if you are looking at a very dim object (like a remote galaxy through a telescope's eyepiece). Sometimes the object in your telescope is invisible if you look directly at it, but will suddenly flash into view if you look off to the side where the more sensitive rods pick up the image.

You can also use the fact that through evolution, your brain has developed the capability of better seeing objects that are in motion than those that are stationary. If a fly is sitting motionless on the arm of your couch, you may not know that it is there, but as soon as it takes flight, it may instantly catch your attention. You'll notice this phenomenon if you are looking at an object that is just above the threshold of visibility. It may fade to invisibility if you stare through the eyepiece, but if you tap the telescope with your finger to make the image jiggle, it'll snap back into view. If either of these two techniques fail initially, you can even find any sizable piece of dark cloth to shroud your head while peering through the eyepiece, or simply cup your hands around the eye you are using. This will block any stray light that may be coming from nearby streetlights or from the sky glow that permeates the sky from nearby cities or even the stars themselves.

While the best environment in which to be when you are star-gazing is total darkness, you might want a flashlight so it won't be necessary to root around blindly for your coffee thermos or in your accessory case to switch eyepieces. While any ordinary flashlight emitting white light will damage your night vision, red light won't. Red astronomer's flashlights can either be purchased from astronomy stores, or you can put a small sheet of transparent red plastic across the face of any ordinary flashlight. These points of red light will be seen floating around the participants of the star parties of every community's astronomy club. Other optional accessories might include filters that screw onto your eyepieces or slide onto the front of the telescope tube. These filter out unwanted wavelengths that might reduce detail or contrast, thus enhancing features that you want to see. Examples of filters include oxygen III filters for viewing emission nebulae, moon filters to cut down glare when viewing the moon through a large aperture telescope, solar filters for viewing surface features or prominences on the sun, colored filters to bring out certain details on planets, and light pollution filters that block the typical wavelengths given off by sodium and mercury vapor lamps that illuminate most city streets and buildings.

TELESCOPES AND MOUNTS

Equipped with this information about what is up there in the night sky and how to view them, you'll feel more comfortable as you begin to explore the stars, constellations, and even the universe as a whole. But if you want to buy a telescope, you'll need to know what kinds there are and which one suits you. There are three major types of telescopes: refractor, reflector, and catadioptric. A refractor utilizes lenses to focus light into the eyepiece and has the classic long, skinny shape that most people envision a telescope to have. These telescopes give the sharpest images and are good for medium to high power views of planets, double stars, and star clusters. Reflector telescopes use mirrors to focus light and are more suitable for low power views of dimmer and diffuse deep sky objects, like galaxies and nebulae. Catadioptric telescopes can be considered as hybrids of the two and utilize both lenses and mirrors to focus light. An example of a reflector telescope would be what is called a Newtonian. This type uses a primary mirror at one end of the telescope tube to focus light back up to a diagonal mirror, and this in turn sends the light through the side of the tube and into the eyepiece. Another type of reflector is called a Cassegrain. This type also uses two mirrors but sends the light three times through the telescope tube before it enters the eyepiece at the far end of the tube. The advantage of Cassegrains is that they have shorter tubes than Newtonians of similar aperture, making them easier to store and transport.

Once you have a telescope, you'll need to join it to a mount. Newtonians can be placed either on an equatorial mount (allowing accurate tracking of an object as the Earth rotates), or an inexpensive Dobsonian mount (an altitude-azimuth type of mount). Cassegrains can be placed on a fork mount, either oriented equatorially or in an alt-azimuth position, or on an equatorial mount. The key advantage of an equatorial mount, which holds the telescope in an orientation corresponding to the Earth's rotational axis, is that once it is polar aligned (as explained later), it will accurately track an object whether manually or automatically by a built in drive motor as the Earth rotates. An object's movement can be quite pronounced when using high power as it will zip through your field of view in a matter of seconds. An equatorial mount will also enable pointing the telescope to a specific object by using the mount's setting circles (displaying the RA and Declination where the telescope is pointing) without having to look through a finder scope or the telescope's main optics.

The most important aspect of a telescope other than the quality of its optics is not its magnification, but its aperture (or how large a diameter it has to gather light). If you think of a telescope as a funnel that collects light instead of liquid to send through the eyepiece, then you'll realize the importance of a large aperture. The bigger the aperture, the more light is captured to be focused through the eyepiece, so you'll be able to see dimmer objects and with better resolution as well. Small refractors can be purchased at a modest price, but for most of us, they can get prohibitively expensive when you get to large telescopes. Reflector telescopes can be bought at more reasonable prices even for large instruments. A refractor of any size will allow you to see any of the bright objects such as the moon, planets, star clusters, and brighter nebulae. A reflector in the 4-inch to 8-inch size range will bring most of the dimmer nebulae and galaxies into view. Telescopes between 8 and 11 inches will reach some of the 12th and 13th magnitude objects. And finally, telescopes larger than 11 inches will capture almost anything that holds interest for the amateur astronomer.

As for prices, reasonably sized refractors can retail for less than \$200 to more than \$2,000, and reflectors can range between \$150 to \$5,000 depending on size, what kind of mount is used, and other features such as computerized tracking drives. As a rule, the larger the aperture, the larger the cost. Also, reflectors come in larger apertures than refractors and are less expensive per inch of aperture than their counterparts. If you are trying to decide on what kind of telescope to buy, consider their cost, size and weight (if too large or heavy, will that discourage you from setting it up and using it?), and where you'll usually be observing (a large aperture instrument has less benefit in light polluted cities where most of the dim deep sky objects will be obscured). Consider as well whether you want your telescope to have a motor driven equatorial mount or a computer that will point the telescope automatically to just about any object you want to view. Computerized telescopes will enable you to select an object on a handheld entry pad, and then watch as the telescope automatically points itself to it. They are very prevalent from certain telescope manufacturers, notably Celestron and Meade. You'll have to decide what kind of observer you'll want to be: one who wants the instant gratification of a computerized, automatic go-to telescope, or one who likes the "hands-on" experience and satisfaction of finding objects yourself. The general guidelines below will help you decide on the variety of instrument to buy if you are considering purchasing a telescope.

Small refractor (60-100 mm aperture): good for stars, star clusters, planets and bright deep sky objects. It is appropriate for light-polluted city skies and a small budget. This type of telescope is very easy to transport and maintain.

Small reflector (4"-6" aperture): good for moderately dim deep sky objects of magnitude 8 to 10 in dark skies.

Medium reflector or catadioptric (8"-11" aperture): good for very dim deep sky objects of magnitude 10 to 12 in dark skies and moderate budgets. Cassegrain telescopes are easier to store and transport because of their compact design, but are more expensive than the Newtonian variety of reflectors.

Large reflector (12"-24" aperture or more): good for very elusive objects dimmer than magnitude 13 in very dark skies when cost and handling a very large telescope is not a concern.

FINDING OBJECTS

So, once you have obtained a telescope and have set it up, how do you find objects? If you don't have an automatic go-to telescope that will find objects for you, and the object is too dim to easily see in your finder scope, there are three methods of doing it yourself: star hopping, using setting circles to dial in an object's coordinates, and using a non-magnifying LED finder.

First, let's discuss star hopping. This is simple if you know where an object is by locating it on a star chart. Then you can be aided by "hopping" from one bright star to the next as if following a trail to your target. Star hopping might go something like this: you start by locating a bright, naked-eye star in your eyepiece or finder scope. There is another fairly bright star you can easily find two degrees west, so you pan over to it. Another recognizable star is then one degree south from there, and then your object, perhaps a 12th magnitude galaxy, is one final half degree “hop” to the southwest. There you’ll have it! If you can plan a pathway such as this from any bright starting point, then everything can be located in this fashion with more or less ease.

The second method takes a little more preparation because it requires you to polar align your scope. This will align your telescope with the Earth’s rotational axes so that the movement of your telescope will follow precise lines of RA and Declination. Accurate alignment is desirable, but perfect alignment is not necessary for our purposes here and is reserved only for long exposure photography. First, your mount will have a latitude gauge, and you’ll need to set this for the latitude of your location. The next step is to orient the mount roughly in a north-south direction. Since Polaris, the north star, is only about 3/4 of a degree away from the north celestial pole, centering Polaris in the main optics of the telescope will align the mount with sufficient accuracy for casual observing. Or for more precise alignment, offset your telescope from Polaris 3/4 degree toward the bowl of the Little Dipper and you’ll be almost perfectly aligned with the pole. If your latitude adjustment is correctly set and Polaris is in your eyepiece, then your declination circle will read 90°. Your north-south axis as well as your east-west axis should now be fairly accurate. Now to find an object, first point your telescope to a nearby star whose RA and Declination are known to you. When that star is in your eyepiece, your Declination circle will read the correct value for it. Adjust the moveable RA circle to the correct value as well. When you move the telescope so that the circles read the new RA and Declination of the object you are looking for, it should then be in or near the field of view of your eyepiece, depending on the accuracy of your initial alignment.

Finally, an LED finder attaches to your scope in addition to, or in place of, your regular finder scope and uses an LED point of light or target circles (in the case of a Telrad brand finder) projected on a transparent screen at the front of the finder. This device makes what looks like a red image projected onto the sky, and the method of using it is to use both eyes to see where the image is superimposed among the stars. Once you have moved your telescope so that the red image is directly over the spot in a constellation where an object lies, then it should be in, or sufficiently near, the field of view of the main optics of your telescope.

FINAL NOTES

If there is a local astronomy club in your area, then you'll benefit from becoming a member. This will enable to you participate in the club's star parties at their dedicated observing site and explore other activities to nurture your interest in astronomy. Observing in a group is both more rewarding and safer than observing alone if you are out at a remote, dark site. If going solo, it's important to carefully choose the location at which to set up your telescope so that you aren't trespassing and won't be bothered by intruders (human or otherwise). On several occasions when it's been just me and my telescope out in the desert or on a dead-end street on the outskirts of the city, I have been questioned by the police, wondered if those behind the approaching headlights were friendly, and been investigated by coyotes and various other unidentified wild animals making sounds in the bushes. Once I shined my flashlight out into the darkness toward a noise only to see a pair of eyes reflecting back at me! Luckily, whatever it was moved on.

On the other hand, setting up in a public place (such as a park or even on the sidewalk in front of where you live) can be rewarding as well. The innate curiosity of strangers will prompt them to approach you and ask what you are looking at, and you can offer them a look. While the average person may not have any appreciation for seeing a dim nebula or galaxy, he or she often will be amazed at views of recognizable objects such as Saturn, the Moon, and even some of the bright star clusters.

With all of this said, happy observing!

I have loved the stars too fondly to be fearful of the night.

--Sarah Williams (1837-1868)
British poet

The treasures hidden in the heavens are so rich that the human mind shall never be lacking in fresh nourishment.

--Johannes Kepler (1571-1630)
German mathematician and astronomer

Joy in looking and comprehending is nature's most beautiful gift.

--Albert Einstein (1879-1955)
Theoretical physicist

PART TWO

THE SKIES

Each of the four seasons of the year have their own unique set of constellations that adorn the sky in the early to middle evening hours. If you know the sky well enough, it is possible to roughly determine the date if you know what time it is, or vice versa, just by recognizing the visible constellations and the position of the celestial sphere as it rotates across the sky. Since there are 24 hours in a day and 12 months in a year, the stars traverse the same distance in two hours as they do when comparing how the sky looks at the same time of night from one month to the following month. In other words, you'd have to look at the sky two hours earlier for each month that passes for the constellations to be in the same positions as the previous month. After an entire season passes, the constellations which had been in the sky are then behind the sun (or in the early morning sky) and replaced with a whole new set of evening constellations. The following will give you a quick tour through the sky during each of the four seasons and point out the constellations that appear during these periods based on the observer being outside at 10 pm in the middle of each season. Also refer to the sky charts included elsewhere in this book for the shapes and placements of the constellations relative to each other, as well as for objects that might be of interest.

During the winter months of December, January, and February, several of the brightest stars visible from Earth show themselves. The "Winter Hexagon" asterism is made of some of the most prominent stars of the constellations Taurus, Auriga, Gemini, Canis Minor, Canis Major, and Orion (who stands within the hexagon), and covers over a quarter of the sky from near the zenith to over halfway to the horizon. Closer to the pole you'll find Ursa Major with it's Big Dipper asterism rising in the East. The same distance on the opposite side of the celestial pole is Cassiopeia and Cepheus, the king and queen. Also nearby are Perseus and Andromeda, with Pegasus beginning his descent below the horizon in the west. Low in the southwest is Cetus, the whale (or sea monster). To the southeast of Cetus are the dim stars of Fornax and Eridanus. Rising in the East late in the season are Cancer, Leo, and the head of Hydra. Also sharing the winter sky are the dimmer constellations of Monoceros (near Orion), Lynx (near the nose of Ursa Major), and Camelopardalis (east of the bright stars of Cassiopeia).

The warmer months of Spring present both the Little Dipper (Ursa Minor) and the Big Dipper (of Ursa Major) high in the sky. Following the curve of the Big Dipper's handle will bring you to a bright, orange star called Arcturus in the constellation of Boötes, and continuing this curve takes you to Spica, the brightest star in the constellation of Virgo. South of Ursa Major's Big Dipper is Leo, and south of Leo is Crater, Corvus, and the winding body of Hydra (which covers about half the distance between the southwest and southeast horizon). Low in the east beyond the tail of Hydra is Libra, and just crawling above the horizon is Scorpius with its bright star Antares. The dim constellations of spring include Leo Minor (directly north of Leo), Coma Berenices (sandwiched between Leo and Boötes), Canes Venatici (below the Big Dipper's handle) and Sextans (south of Leo's head). Draco (wrapping around Ursa Minor) and Hercules are in the East. Nearby, next to Boötes, is Corona Borealis (the northern crown), and below this is Serpens Cauda (the head of the snake). The spring sky is famous for the dozens of galaxies that inhabit the Virgo-Coma cluster. As the name suggests, this cluster lies across the border of Virgo and Coma Berenices.

Summer brings the brightest part of the Milky Way into view. Its bright band slices across the sky from north to southeast, and later in the season, from northeast to southwest. Because of this fact, the summer sky is known for vast numbers of star clusters and nebulae, most notably in Sagittarius and Scorpius. The Big Dipper stands vertically, and Ursa Minor, or the Little Dipper, is pointing straight up toward the zenith. Just south of Ursa Minor is Draco, where it is poised best for viewing. In the summer sky is another geometrical shape, the Summer Triangle. This is made by the three brightest stars in the constellations of Lyra, Cygnus, and Aquila. Hercules reaches his highest point. The far southern constellations of Lupus and Centaurus slide above the horizon for a short time. Also, several small constellations appear during the summer, and are notable for lying along the Milky Way. These include Vulpecula, Sagitta, Delphinus, and Scutum. Between Hercules and Scorpius is Ophiuchus, with Serpens Caput and Serpens Cauda making the snake that is sliding across his body. Boötes and Corona Borealis are in their prime near the meridian.

Autumn gives us the zodiacal constellations of Capricornus, Aquarius, Pisces, and Aries. The Great Square of Pegasus is prominent close to the zenith, with Andromeda, and its bright naked eye galaxy, branching off toward the northeast. Cetus, with its many galaxies, is rising, and the Summer Triangle is still visible before it sets in the west. Most of the stars of fall are dim, but Piscis Austrinus has its bright, first magnitude star called Fomalhaut sitting all by itself in the south. To the north is Perseus, displayed right next to the W shaped constellation of Cassiopeia, and which is well-placed with the queen to be in view all night long. The less known constellations of fall include Lacerta (north of the legs of Pegasus), Equuleus (just in front of the nose of Pegasus), Microscopium (below Capricornus), Sculptor (east of Piscis Austrinus), and Triangulum (fitting in between Andromeda and

Aries). Several of the fall constellations mentioned above make this the "wet" part of the sky, since many of them are connected to water.

THE CONSTELLATIONS

In all, there are 88 constellations in the entire celestial sphere that envelopes the Earth. Some of these, however, are eternally below the horizon for those of us with vantage points at northern latitudes. This leaves us with the 64 constellations described below. The dates given for best observing them occur when the constellations are near the meridian within 6 hours after darkness. However, if you are observing before dawn, they can be viewed up to 2 or 3 months prior to the earlier date that is indicated. The sky charts that appear elsewhere in this book provide you with a basic map of where the constellations appear for each season, with the constellations labeled using their abbreviated names. They represent how the sky looks for the date and time indicated if you were to lie on your back with your feet pointing south. This will give you a general layout of the constellations. However, if you plan to use a telescope for searching the deep sky objects that are contained within each constellation, then it would be beneficial to purchase a good star chart that will show the sky in a more detailed, larger format.

Andromeda (an drom' mi duh)
Abbreviation: And
The Chained Maiden
Best observed between September 15th and January 15th

Andromeda is one of the oldest constellations, probably dating back to the 5th century BC. The stars of Andromeda trace out a vague shape, but its depictions are of a young woman chained to the rocks beside the sea. She found herself in this position as a result of events that began with the boastfulness of her mother, the queen Cassiopeia. Along with her husband, king Cepheus, Cassiopeia ruled an ancient land on the coast of Africa, and she was known for frequently declaring her beauty. The gods strongly disapproved of this boastfulness, and when they could tolerate it no more, they had the sea god Poseidon send Cetus (a sea monster) to terrorize the coastline. In desperation, Cassiopeia offered her daughter, Andromeda, as a sacrifice to Cetus in the hopes that it would send the monster away. Securing Andromeda to the rocks at the water's edge indeed attracted the monster's attention, but her screams also summoned Perseus, the heroic son of Zeus who was passing by and who successfully rescued the princess. More details of the events and characters of this story can be found in the description of other constellations that play a role, including Pegasus, Cassiopeia, Cepheus, and Perseus himself.

Andromeda can be easily found as fall arrives. The main pattern of stars is two strands branching out from the northeast corner of the great square of Pegasus south of Cassiopeia. It contains stars of 2nd and 3rd magnitudes.

A deep sky object of unrivaled prominence among all the others in the northern hemisphere is found in Andromeda. Even while sitting 2.2 million light years away, it is the most distant object that most people can claim as visible to the naked eye. Due to its size and proximity, it covers an area of sky 3 degrees across (or 6 times the diameter of the full moon) although varying amounts of this expanse will be seen depending on the darkness of your sky. What we have here is the Andromeda Galaxy, or M31 (the 31st object cataloged by the French astronomer Charles Messier). This is a spiral galaxy that is similar in size to the Milky Way, both of which are members of what is called the Local Group. This is a collection of about 30 galaxies of various sizes forming a small cluster occupying a parcel of space roughly 2.5 million light years across. To find M31, start with Beta (β) Andromedae and follow a line through Mu (μ) and Nu (ν) to its unmistakable magnitude 4 glow. Or, you can use nearby Cassiopeia to point the way. The taller triangle of Cassiopeia formed by Alpha (α), Beta (β), and Gamma (γ) Cassiopeiae points almost directly to M31, using α as the pointer star. Through a telescope, M31 shows a very bright central core, and because of the galaxy's slanted orientation to our line of sight, its tight arms overlap each other and fade as a single mass to the edge of this large system. Any significant magnification will make the galaxy spill over the field of view of your eyepiece, requiring you to pan your scope from side to side to see it all. Large scopes will bring out details in the dust lanes that parallel the spiral arms and contrast with the mostly uniform glow of the galaxy's 200 billion stars. While you can enjoy the detailed view of M31 through a large, high magnification telescope, perhaps the best views can be obtained through good binoculars or short tube refractors. The low power, in this case, will enable you to see the entire galaxy at once poised behind a scattering of our own Milky Way's stars.

Close by is another Messier object, M32, which is actually a small satellite galaxy of M31. M32 is a small oval spot south of M31, and it can be overlooked if concentrating your view on M31 at high power. Low power brings them both into view and illustrates the relationship between the two. If M32 were by itself, it would still be an attractive telescope target because its nearness makes it large and bright (magnitude 9.1). M31, however, greatly overshadows it.

Apparently, Messier returned to this area later to catalog M110, another satellite galaxy of M31. This one is almost double the size of M32, but only a little brighter at magnitude 8.9, so therefore somewhat less striking due to the lower surface brightness. This system sits a little more than half a degree to the northwest of M31's center.

Elsewhere in the constellation you'll find a beautiful double star in Gamma (γ) Andromedae, magnitude 2.1. This pair of stars consists of a bright primary orange-yellow K type star of magnitude 2.2 and a blue magnitude 5 secondary star of spectral type B. They are separated by about 10 arc seconds. This is truly considered one of the most beautiful double stars in the sky.

Three and a half degrees almost due east from γ is NGC891, a fairly bright edge-on spiral galaxy of magnitude 10.8 and 13.4' x 2.5' in size. In medium-sized telescopes, you'll see an unmistakable spear of light a little less than a full moon's diameter. It's exact location is RA 2h 22.5m Dec +42° 21', or 1 degree due north of the 5.8 magnitude star SAO 37986.

Staying in this vicinity and looking down slightly less than five degrees south of γ will bring us to NGC752, a large open star cluster almost a full degree in diameter. The brightest stars of this cluster are of magnitude 7 and 8, and it is best appreciated through a wide angle telescope or binoculars. High power will immerse you too deep into the cluster and it will lose its identity. NGC752 is centered at RA 1h 58m Dec +37° 41'.

Moving over to the western part of the constellation will bring us to NGC7662. This is a planetary nebula given the whimsical nickname of the Blue Snowball. Indeed, through a telescope, its gasses may have a bluish tint to them. It glows at magnitude 9 over an area of about 30". This nebula sits in a rather empty area of Andromeda, so finding it might be difficult. Use your setting circles or the go-to capabilities of your telescope to center your instrument at RA 23h 26m Dec +42° 33', or start at magnitude 4.3 Iota (ι) Andromedae, pan 2 degrees west to magnitude 5.8 13 Andromedae, then slide not quite a half a degree southwest from there to the nebula.

Aquarius (uh kwair' ee us)
Abbreviation: Aqr
The Water Bearer
Best observed between September 1st and November 15th

This constellation shares a part of the heavens that was seen as a celestial sea in ancient times. Nearby we see other constellations that have ties with water such as Cetus, Pisces, Capricornus, and Delphinus. Aquarius depicts a human figure pouring water from a jar, an act that some cultures considered to be the source of the river Eridanus, or responsible for the global flood of Biblical times. Conversely, Aquarius has also been seen as Zeus pouring life giving waters from the sky.

To spy the rather dim stars of Aquarius, look due south near midnight at the beginning of September, or closer to 8pm at the end of October. Aquarius is between the head of Pegasus to the north and the solitary 1st magnitude star Fomalhaut to the south. One conspicuous feature of Aquarius is the Y-shaped asterism called the "water jar" marked by Eta (η) , Zeta (ζ) , Pi (π) , and Gamma (γ) Aquarii near the center of the constellation.

Aquarius abounds in diverse deep sky objects. The largest, but often the most difficult to see, is NGC7293, called the Helical, or Helix, nebula. The proximity of this planetary nebula (only about 450 light years) curiously adds to its elusiveness. It glows at magnitude 6.5, but that light is spread out into a circle that is 13' across, which is almost half the diameter that the moon covers in our sky. To find it, start at magnitude 3.7 88 Aquarii, then move 35 minutes in RA west to Upsilon (υ) Aquarii, then another 5 minutes of RA, or 1.2 degrees, west to the Helix. You will want to try to observe this nebula in transparent, dark skies in a low power eyepiece capable of at least a 1/2 degree (but preferably more) field of view. This will give you some contrast between the edges of the nebula and the darker background. You will see a large, ghostly circle of light. In large scopes, you may notice that the Helix is slightly ovoid, shows a pronounced darkening near the center, and you will have a chance to see the magnitude 13 central star.

Moving over to the western half of the constellation, we will find three targets within 3 degrees of each other. NGC7009, on the northeast corner of this triangle of objects, is another planetary nebula nicknamed the Saturn Nebula. It earned this name from the spikes that jut from the ends of its oval shape, thus resembling the rings of the planet Saturn. At magnitude 8.3, it is almost two magnitudes dimmer than the Helical Nebula, but it measures a compact 25" across. This means that its higher surface brightness stands out against the background sky. NGC7009 is located at RA 21h 41.2m Dec -11° 22', or 1.3 degrees west of 4.5 magnitude Nu (ν) Aquarii.

Three degrees WSW of NGC7009, or RA 20h 53.5m Dec -12° 32', is globular cluster M72. It measures 6' across and shines at magnitude 9.4. In most telescopes, M72 shows little texture. Its 15th magnitude stars are difficult to resolve, except in the largest aperture instruments, so most views will simply show a soft, round glow.

An interesting object that is 1.3 degrees east of M72 represents the third item in our trio. When Messier studied it in October of 1780, he described it as a cluster of three or four stars in a surrounding faint nebula, therefore he added it as the 73rd object in his catalog. Indeed it may look like a fuzzy spot if seen at low power with averted vision, but closer examination clearly reveals only a tight group of four 10th, 11th, and 12th magnitude stars in a general "Y" shape with no nebulosity. This little "cluster" can be hard to identify unless you know exactly what you are looking for. M73 is located at RA 20h 59m Dec +12° 38'.

For a genuine challenge, you may want to search for globular cluster NGC7492. This is a small, magnitude 11.5 cluster at RA 23h 8.4m Dec -15° 37'. Its stars are extremely faint and sparsely scattered across a small area of 5'. If you start at Delta (δ) Aquarii, drop south one degree, then move east 15' in RA. NGC7492 will then sweep into view.

Very close to the celestial equator is the best globular in Aquarius, M2. It's 12' in diameter and a bright magnitude 6.5. It sits in a rather blank part of the constellation 4.8 degrees north of Beta (β) Aquarii (RA 21h 33.5m Dec -00° 49'), so distinguishing it in your finder scope as a hazy spot is easy. Small telescopes will see a round, fuzzy glow. Medium to large scopes will begin to resolve stars deep within the core of this globular.

Aquila (uh kwi' luh)
Abbreviation: Aql
The Eagle
Best observed between July 1st and August 31st

The constellation of Aquila dates back to about 1200 BC. It was then that early astronomers began seeing it as a bird or, more specifically, the eagle which was something of a companion or servant to the ruler of the gods, Zeus. Aquila held Zeus's thunderbolts until he decided to use them, and the eagle went to Mount Ida on Zeus's orders to kidnap a young shepherd called Ganymede and bring him to Olympus so he could serve as the gods' cup-bearer. In the far east, Altair, the brightest star of the constellation, represented a royal herdsman who began neglecting his duties when he fell in love with a young maiden, herself preserved in the sky as the star Vega. The sun god, who was the maiden's father, took displeasure in this and separated the two by placing them on opposite sides of a mighty river, and this was the reason given to explain why Altair and Vega appear on opposite edges of the "river" that we call the Milky Way. In addition, Indian folklore describes some of Aquila's stars as the footprints of the god Vishnu. It's not surprising that Aquila's distinctive shape and bright stars have produced such stories among these widely separated cultures.

This constellation's individual stars that shine among the star fields of the Milky Way, where Aquila flies, include Altair (an Arabic name meaning "flying eagle"), of spectral type A7 IV-V, magnitude 0.8, and a mere 17 light years away. An interesting characteristic of Altair is its rotational speed. While our own sun takes more than 25 days to complete one rotation (measured at the equator), Altair spends less than 10 hours to do the same. As a result, the star's shape is flattened at the pole due to the stretching caused by centrifugal force. Altair forms one corner of the pattern called the Summer Triangle. (Deneb in Cygnus and Vega in Lyra comprise the other two corners of the triangle.)

To the south lies Eta (η) Aquilae, a member of the class of giant variable stars called Cepheid variables. Its magnitude changes from 3.7 to 4.5 every 7 days and 4 hours, and in the process undergoes a change in spectral type from G2 to F6. Near the tail of the eagle, approximately one degree to the southwest of Lambda (λ) Aquilae, is V Aquilae. This is another variable star ranging between magnitudes 6.6 and 8.4, but what sets this star apart from the others is its deep red color. Its surface temperature is less than 2000 degrees K. This classifies it as a star of spectral type N. Point your telescope in its direction and you should be able to easily distinguish it by its color. It lies at the end of an arc of 7th and 8th magnitude stars curving away from 4th magnitude 12 Aquilae.

Although Aquila does not contain any bright Messier objects, it is, however, host to some deep sky targets that are observable through small telescopes provided you have dark skies. One of these is NGC6709, a magnitude 7.5 open cluster located 3.5 degrees south and 14' west of Zeta (ζ) Aquilae (RA 18h 51m 29s Dec +10° 21' 01"). About 30 to 40 stars reside in this modest cluster. NGC6781, one of the 11 planetary nebulae within Aquila, is more of a challenge and can be found about 3/4 of the way along a line drawn from Zeta (ζ) to Delta (δ) Aquilae (RA 19h 18m 20s Dec +6° 30' 00"). This is one of the larger and brighter planetary nebulae in this region, but still no brighter than 11th magnitude. It appears as a neatly circular smudge of light about 4 minutes of arc in size. A 6" or larger scope is required to make this dim nebula obvious, although it can just be discerned in instruments as small as 80 mm.

Another 11th magnitude object nearby is NGC6760, a globular cluster 2 degrees south and 15' west of Delta (δ) Aquilae (RA 19h 11m 10s Dec +01° 01' 42"). Use magnification of 80x or more to bring this small globular into view.

While these are the highlights of Aquila, you may find many more objects during your own explorations.

Aries (air' eez)
Abbreviation: Ari
The Ram
Best observed between November 1st and January 15th

Many ancient civilizations have viewed this inconspicuous group of stars as a ram. Its pattern suggests the curved shape of a ram's horns and is a fitting tribute to an animal revered by both nomadic peoples and agricultural settlers alike. Specifically, Egyptians of 16 century BC considered this constellation to be the ram connected with their god Amon Ra. In Greece, it was the golden fleeced ram created by the god Hermes to take the son and daughter of King Athamas of Thessaly away from their abusive stepmother. The ram's final destination was a region near the Black Sea where the animal was sacrificed. Its golden fleece was placed in the branches of a tree and guarded by a dragon, but nevertheless it was later stolen by Jason and his Argonauts.

Looking high in the south, you'll find Aries crossing the meridian near 10pm on the 1st of December and tucked between the tall V shape of Pisces and the soft glow of the Pleiades star cluster of Taurus. Its brightest stars range from magnitude 2 to 4.

An excellent starting point for your exploration of Aries is the magnificent double star Gamma (γ) Arietis. To the naked eye, it glows softly at magnitude 3.8. Through the telescope you will see two white points of light oriented north-south and separated by 8 arc seconds. The two stars are twins in most respects. Each are magnitude 4.8 and of similar spectral types: A0 and B9. In the year 1664, Robert Hooke discovered this double star accidentally while comet hunting. It was one of the first double stars to be discovered, and since then, while hardly any change in PA has been observed, its separation has decreased slightly. This indicates that the orbital plane must be turned flat with respect to the Earth and in a few more centuries the two stars may appear to meet each other with very little or no separation at all.

Another nearby double star, and one easier to split from its 37" separation, is Lambda (λ) Arietis. The primary glows at magnitude 4.9, and the secondary at 7.7, making its overall brightness 4.8.

Since Aries is far from the Milky Way's spiral arms, the deep sky objects here are all in the form of galaxies. But they suffer from remoteness. However, well worth tracking down is NGC772, Aries' brightest galaxy. Comparatively strong at magnitude 11 and measuring 7.3' x 4.3', this galaxy shows a distinct core, and in CCD or photographic images, it reveals an assortment of spiral arms. One arm is elongated dramatically as if being pulled by some unseen attractor. NGC772 is 1.4 degrees east southeast of Gamma (γ) Arietis at RA 1h 59.4m Dec +19° 00.4'.

NGC1156 is an irregular galaxy measuring a small 3.3' x 2.4' and a little dimmer at magnitude 12. Large telescopes may see a mottled, rectangular object, or otherwise simply an undefined patch of light of roughly oval shape between two dim stars. You'll have to look to the northeast part of the constellation 3 degrees southeast of 41 Arietis, or RA 2h 59.6m Dec +25° 14.3'.

Auriga (are rye' guh)
Abbreviation: Aur
The Charioteer
Best observed between December 1st and February 28th

Auriga is mainly considered to be a charioteer, but it also has a dual role as a goat herder. In some depictions it appears as a man on a chariot, but carrying a goat and three kids in his arms. The identity of Auriga is said to be either the Greek character of Hephaestus, or his son, Erechtheus. These two were both crippled, and thus came to invent the chariot to aid their mobility.

Auriga is easily found if you look to the northeast in the evening during late autumn. It forms the shape of a pentagon [with its brightest star, called Capella, and Beta (β) Tauri (formerly Gamma (γ) Aurigae) forming two of the corners opposite each other] north of Orion and sandwiched between Gemini and Perseus.

In the late fall and early winter, Auriga is above the horizon before the sun sets. By virtue of its northerly declination, it remains in the sky all night as it makes a slow march toward the west while the constellations farther to the south reach the obscuring earth sooner. Acting as a regional beacon is the sixth brightest star in the sky, Capella. This is a star that is similar to the sun in color and temperature (a spectral type of G8 for Capella and G2 for the sun), but it is a giant star producing the light of 160 suns and thus shining at magnitude +0.1 from a distance of 42 light years. Capella approximates the appearance of our sun if we could view it from a distance of only a few light years.

Auriga is immersed in the Milky Way, so we don't have to look long for some excellent star clusters. Charles Messier cataloged three in his studies of the constellation: M36, M37, and M38. M37 (mag. 6.2) sits just outside of the pentagon shape and is sure to be a favorite. It is the largest, brightest, and richest of the three, containing over 100 stars near 12th magnitude and brighter, with perhaps 500 stars overall. It rivals the appearance of a loose globular cluster with its dense population of stars. If you draw a line one-third the distance between Theta (θ) Aurigae and Beta (β) Tauri, and then move not quite two degrees southeast you will easily spot M37 (RA 5h 53m Dec +32° 33').

Using the same line connecting θ Aurigae and β Tauri will enable us to find M36 (mag 6.5). This time, look half way between the two stars and about the same distance to the northwest as we used for M37. M36 (RA 5h 36.3m Dec +34° 08') contains about 60 young stars, the brightest of which are all of spectral class B2 through B8. This means that they have hot surface temperatures near 20,000 degrees Kelvin. This cluster is the smallest of the trio, but shining slightly brighter than our next target that sits 2.3 degrees to the northwest. This is M38 (mag. 7.0; RA 5h 28.7m Dec +35° 50'), a cluster also boasting many hot A and B type stars with some G-type giants interspersed. When looking at M38, you should also spot a small magnitude 8.2 cluster half a degree to the southwest. This is NGC1907. Measuring 7' in diameter, it is less than half the size of M38.

Near the Psi (ψ) complex of stars is NGC2281. This star cluster measures about 15' across and presents itself as a semicircle or bowl shaped collection of magnitude 7 and dimmer stars. Look for NGC2281 at RA 6h 49.3m Dec +41° 04', or about 3/4 of a degree to the SSW of ψ7 Aurigae.

For a real challenge, try IC2149. This is a tiny planetary nebula measuring 9" across and glowing weakly at 11th magnitude. While actually spotting the nebula is difficult, its location is easy to pin down. It is about 2/3 of a degree west of Pi (π) Aurigae, which in turn is just north of Beta (β) Aurigae, the NE corner of the pentagon. More precisely, NGC2149 is located at RA 5h 56.3m Dec +46° 07.5'. With sufficient aperture and good seeing, you may have a shot at spying its 14th magnitude central star.

To the area southwest of Capella is a group of three stars, Epsilon (ε), Eta (η), and Zeta (ζ), comprising an asterism called "the kids." They are in the shape of a tall isosceles triangle, and the star at the point, ε, is of special interest in part due to the fact that it is an eclipsing binary star of spectral type F0 with an exceptionally long period of 27.06 years, with each phase of deepest eclipse lasting an entire year by itself. Furthermore, its distance is estimated to be near 2,040 light years, so to shine at magnitude 3 (3.8 in eclipse) to an observer on Earth indicates that it is a very luminous star, producing perhaps 20,000 times the light of the sun.

Boötes (boh oh' teez)
Abbreviation: Boo
The Herdsman
Best observed between April 15th and June 30th

As spring approaches, so too, does the bright star Arcturus toward the evening sky. It is not only the brightest star in Boötes, but it's also the brightest by far in its neighborhood of the sky at magnitude -0.04. Being a giant K3 star, its orange color is readily apparent. During June evenings, Arcturus is high overhead on the meridian, marking a spot along the same path if you follow the curve of the handle of the Big Dipper. The general shape of the entire constellation reminds some of us of an ice cream cone or a snow cone (a welcome sight after the hot days of this time of year).

Several mythological stories refer to the character drawn in the stars of Boötes. The name itself comes from the Greek word for "herdsman", and arktouros, from which we get the name Arcturus, is Greek for "guardian of the bear." It is said that Boötes is the son of Zeus and Callisto. After Zeus's wife, Hera, went into a jealous rage and turned Callisto into a bear, Boötes encountered the bear and threatened to kill it before Zeus took the bear away and placed it safely in the sky. Boötes is now forever chasing the bear (Ursa Major) around the north celestial pole.

The Egyptians considered the stars near the pole, meaning that they never set below the horizon, to be evil. Thus, they invented the constellation Boötes (in their eyes, a hippopotamus) to keep these stars confined. The Arabs, by contrast, saw the circumpolar stars as a peaceful flock, again herded by Boötes.

Arcturus, or Alpha (α) Boötis, shows its orange color very well through a telescope or naked eye. It is cooler than our sun with a surface temperature of 4,200 degrees Kelvin. However, its 4 solar masses occupy a volume much larger than our own star, making its stellar density over 3,000 times less. Its large proper motion of 2.29" per year is due to two factors. One is its proximity, but more so because of its motion relative to ours. While the sun revolves around the galactic core within the plane of the galaxy, Arcturus moves within the great spheric-al halo surrounding this plane. Arcturus is currently slicing through the part of the galactic disk occupied by the sun, and in a short half million years, it will have receded in the distance beyond naked eye visibility.

Scanning Boötes with binoculars or a telescope will show you something obvious. You'll see scattered stars with hardly any hint of diffuse objects like clusters or nebulae. But looking more closely at these stars reveals Boötes' main attraction: it is a true playground for fans of double stars. The following is a list of the best of Boötes's doubles:

Upsilon (μ) Boötis (Alkalurops): the easiest double star with a wide separation of 109" between its primary (shining at magnitude 4.5), and its secondary at 6.5. If you really increase the power, you may be able to split the dimmer component that is separated by a scant 2".

Iota (ι): a similarly easy separation of 38.5". Its contrasting brightness of magnitudes 5 and 7.5 make for an interesting pair.

Kappa (κ): a bit over half a degree to the northwest of ι. Here is a star of magnitude 4.5 next to a 6.5 companion with 13.3" of separation.

Pi (π): down in the vicinity of Arcturus is this pretty blue-white double. The magnitude 5 and 6 pair are separated by 5.6".

Xi (ξ): jump 3.7 degrees northeast of π to this fairly easy double. This one takes on the appearance of a yellow or gold magnitude 5 primary star of spectral type G8 paired with a reddish magnitude 7 star of type K4. The separation of 6.9" makes it easy to split in small scopes at high power.

39: in the northern reaches of the constellation is a close 6 and 6.5 magnitude pair. Only 2.9" separate this white couple.

Epsilon (ε): called "Pulcherrima" (Latin for "most beautiful") by its discoverer F.G.W. Struve in 1829. This is an attractive orange-blue pair. It is easy to find, but difficult to split. The magnitude 2.7 primary is 2.6" from magnitude 5 secondary.

Zeta (ζ): the benchmark some observers use to gauge the culmination of their equipment's capabilities, their skill as observers, and steadiness of the skies. The magnitude 4.5 and 5 pair is separated by only 0.9".

Out of the 20 or so galaxies found on a detailed star map within Boötes, NGC5248 is the brightest. Its magnitude 10.7 glow reveals an oval measuring 6.1' x 4.4'. This object is located at RA 13h 37m 32s Dec +08° 53' 08" in the far southwest corner of the constellation. Look near Boötes's boundary with Canes Venatici to find NGC5466, a loose globular cluster of a fairly bright magnitude 9.1, and having a sizable diameter of 11'. Its location is RA 14h 05m 30s Dec +28° 32' 00".

Camelopardalis (ka mel oh par' dahl iss)
Abbreviation: Cam
The Camel Leopard, or Giraffe
Best observed between December 1st and March 1st

One of the more challenging constellations, both to find and to pronounce, is nestled among Auriga, Perseus, Cassiopeia, Ursa Major, Draco, and Ursa Minor. If you make an imaginary triangle with Capella, Alpha Persei, and Polaris, the brightest stars of Camelopardalis will stretch across it. The rest of the constellation spans westward to the nose of the Great Bear and the tip of Draco's tail. You'll notice that this group of stars is not conspicuous at all. The brightest is only of magnitude 4. The constellation was established by the German astronomer Jacob Bartsch in 1624 to fill in a bland area of the northern sky and was named after the giraffe, or "camel-leopard," which is what the Greeks called the animal that appeared to have the head of a camel and the spots of a leopard. On some star maps, Camelopardus, the older version of this constellation's name, is indicated.

Navigating through Camelopardalis by star hopping could be difficult due to the scarcity of bright stars, but there are two bright deep sky objects that should still be easy to find. The first is NGC1502. This is a sparse open star cluster of magnitude 5.5 located at RA 4h 7.5m Dec +62° 18'. This cluster sits roughly in the center of the five brightest stars that make up the main shape of the constellation. You can find it by moving 56m west in RA and 2 degrees north in declination from Beta (β) Camelopardi. It contains about 25 stars having a wide range of brightness plus two notable double stars.

If you have a keen set of eyes and a sizable telescope, try searching for NGC1501 only 1.5 degrees directly south of NGC1502. This is a magnitude 12 planetary nebula about 50" in size with a central star shining weakly at magnitude 13.5. Some reports give this nebula a bluish tint with uneven texture across its face. It is found at RA 4h 7m Dec +60° 54'.

Toward the east is a striking spiral galaxy, NGC2403. This resembles M33 in Triangulum, except that it is only about 1/15 the size. At magnitude 9, this galaxy is an easy target through any telescope, and can even appear in binoculars as a soft oval smudge. Through a telescope, its nearly face-on orientation is evident and some mottling across its broad arms may be seen in larger instruments. Two foreground stars make an interesting sight on opposite flanks of the galaxy, giving the impression of dual supernovae occurring within the galaxy's outskirts.

Cancer (kan' ser)
Abbreviation: Cnc
The Crab
Best observed between February 1st and April 15th

If it were not for two attributes, its membership among the 12 signs of the zodiac and possession of a premier star cluster, Cancer may otherwise be a little noticed constellation. Its dim stars rise after the winter heavyweights Orion, Gemini, and Auriga, and before those heralding spring like Leo and Boötes. Cancer is found between Gemini (with its twin stars of Castor and Pollux) and Leo. Another way to find Cancer is to start at Sirius (the brightest star of Canis Major, as well as the brightest star in the sky) and extend a line going northeast through Procyon (the brightest star in Canis Minor) and continue about the same distance between the two stars to a rather blank area of the sky.

The legend of the crab goes back several hundred years. A crab was said to have followed Hera's orders to distract Hercules while he was battling Hydra, the sea serpent, but the poor crustacean was killed during the attempt. It nevertheless was awarded a place in the sky, although with dim stars because of its failure to fulfill the assignment. In ancient times, the place in the sky that it occupies was prominent as it contained the point where the sun reached its most northern spot before moving south again, or backwards like a crab, on the summer solstice. Because of precession, that spot is now near the border of Gemini and Taurus, but we still keep the name of the circle of 23.5 degrees north latitude over which the sun hangs on the solstice as the Tropic of Cancer.

In a dark sky, the star cluster M44, also called Praesepe or the Beehive cluster, can be seen with the naked eye as a fuzzy patch of light in the center of the constellation. This is one of the largest and brightest clusters visible from Earth. It glows with a magnitude of 4.5 and covers an area of 80', or almost 3 times the diameter of the moon. It is best viewed with binoculars or a wide field telescope. Ancient stargazers, notably Hipparchus and Aratus in the 2nd and 3rd century BC, made written observations of Praesepe that described it as "cloudy" or "misty." Some thought that it was a gateway between Heaven and Earth where souls descended before being born. It wasn't until Galileo pointed his telescope toward it that this "cloud" was discovered to be a collection of individual stars. Modern observations show about 200 member stars with magnitudes ranging from 6.3 to 14.

A second star cluster is M67, of magnitude 7.4 and about 25' in diameter. This is a breathtaking cluster of a few hundred stars at RA 8h 51m Dec +11° 48' or about 8' west in RA of Alpha (α) Cancri. Unlike most open clusters that are mainly located along the galactic plane, M67 is displaced by almost 1500 light years. Spectra and magnitude studies of

M67 show that its stars exhibit characteristics shared by those in many globular clusters and suggest that this cluster may be as old as 10 billion years; however, stars here are richer in metals. This is more typical of sun-like stars than for the old stars found in the average globular cluster.

There are about half a dozen galaxies in Cancer, but most are too dim for most amateur telescopes. The brightest one, which should be within reach of at least medium sized telescopes, is NGC2775, a magnitude 11.5 spiral located at RA 9h 10.2m Dec +7° 01' (or 4.8 degrees south and 12 minutes east of α Cancri). It's a small oval smudge measuring about 2.2' by 1.5'.

Cancer also offers some pleasing double stars of varying difficulty. The best is Iota (ι) Cancri, marking one of the crab's right legs (or claw). This star is composed of a yellow G8 and a blue A3 pair separated by 30.5 arc seconds and a nearly fixed PA of 307 degrees. Low power is all that is necessary to split this double. A much closer pair is Phi2 (φ2) Cancri. This is a closely matched bluish white A3 and A4 pair separated by 5.1", and a PA of 217 degrees

The last star to point out, in this case a triple star, is Zeta (ζ) Cancri. The wider pair of yellow-orange stars is separated by 5.8" and a PA of 77 degrees, but if you have the aperture (perhaps 12 or more inches), the optics, and the steady seeing required, try separating the closer pair at a mere 0.9". At Zeta's distance from Earth, this is similar to the separation of the Sun and Uranus.

Though Cancer is not marked by bright stars, it is well worth looking at the treasures it has to offer.

Canes Venatici (kay' neez ve nat' e see)
Abbreviation: CVn
Hunting Dogs
Best observed between March 15th and June 15th

Small and marked by only two noticeable stars, this constellation is a fairly modern one created by Johannes Hevelius, a German-Polish astronomer, in the late 17th century from some stars that lie between the handle of the Big Dipper and the constellation of Boötes, the herdsman. In fact, Canes Venatici are Boötes's hunting dogs, often considered to be a pair of greyhounds that has joined Boötes in chasing the two bears, Ursa Major and Ursa Minor, around the celestial pole. Canes Venatici lies directly below the tail of Ursa Major, also known as the handle of the Big Dipper.

Alpha (α) Canum Venaticorum is the brightest star of the

constellation and has the interesting name of Cor Caroli, or "Charles' heart." This name was suggested to British astronomer Edmond Halley by court physician Sir Charles Scarborough as a tribute to King Charles II. Looking at it with a telescope will reveal that it is a double star, one component of spectral type A0 V and magnitude 2.9, and the other, 20 arc seconds away, of spectral type F0 V and magnitude 5.6. The color contrast of the two is very small, but descriptions of the dimmer star range from white, to lilac, to coppery. You may want to decide for yourself what color the two stars exhibit.

A very well-known Messier object is located in the northeast corner of the constellation. You can find it by using Eta (η) Ursae Majoris, the star marking the tip of the handle of the Big Dipper, as your guide. Move slightly more than 2 degrees south and 17 minutes in RA west of this star and you will see an 8th magnitude glow 10 arc minutes in diameter known as M51, the Whirlpool Galaxy. Its coordinates are RA 13h 29.9m Dec +47° 12'. This was the first galaxy to be noticed as showing a spiral structure. Previous accounts describe the galaxy as having a bright central region surrounded by a "ring." A 60 mm refractor will see the unmistakable nucleus surrounded by a faint glow, and instruments 8" and larger will begin to show the photogenic spiral structure and perhaps some detail in the satellite galaxy NGC5195.

Moving to the southeastern edge of the constellation will bring us to M3, one of the brightest (mag. 6) globular clusters in the sky. Although it is located in a rather empty part of the sky, it is easy to find since it lies exactly on a straight line between Arcturus and α Canum, not quite half way from Arcturus. The coordinates are RA 13h 42.2m Dec +28° 23'. This cluster rivals the great Hercules cluster (M13) in size and brightness, and medium-sized scopes will bring out many hundreds of stars of magnitude 11 and fainter.

Returning to the wealth of galaxies in Canes Venatici, our next visit is to M94, a tightly wound bright spiral of magnitude 8.9 that is found at RA 12h 50.9m Dec +41° 07'. M94 forms a triangle with Alpha (α) and Beta (β) Canum directly between and 1.6 degrees above a line connecting the two. Its bright nucleus fades to a soft glow about 4' x 5' in size. M63 is another spiral at RA 13h 15.8m Dec +42° 02', or again using α Canum and Eta (η) Ursae Majoris as markers, move about one-third of the way between the two from α Canum and about 1/2 degree to the east. This 9.8 magnitude galaxy measures 9' x 4' and shows a nucleus surrounded by bright and tightly wound arms which drop in brightness quickly to the outer edges of the galaxy. Our final Messier galaxy here is M106, which is yet another spiral glowing at magnitude 9 and turned slightly from an edge-on orientation. To find M106, look at RA 12h 19.0m Dec +47° 18', or along another line running between β Canum and Gamma (γ) Ursae Majoris, look about half way between the two.

If you can tolerate one more galaxy, find NGC4631, interesting due to its edge-on appearance. It measures 12.5' x 1.2', glows at magnitude 9.7, and may show a mottled appearance in larger scopes due to clumps of stars and dust. (RA 12h 42.4m Dec +32° 35', or 5.8 degrees south and 15' in RA west of α Canum.)

For a colorful treat, look at Y Canum, one of the reddest stars that you'll find in the sky. It is 4 degrees north and about 10 minutes in RA east of β Canum.

Canis Major (kay' niss may' jer)
Abbreviation: CMa
The Great Dog
Best observed between January 1st and March 1st

Before this constellation, southeast of Orion, was known as one of Orion's hunting dogs, it was seen by the ancient Indians as a deer hunter with the stars of Orion representing the prey. The three stars of Orion's belt was an arrow piercing the deer's side. In other times, Canis Major has been considered to be a dog in almost all other prominent stories of mythology and antiquity.

Alpha (α) Canis Majoris probably has more legend behind it than the constellation itself, owing to the fact that it stands alone as the brightest star visible from Earth. Its modern name, Sirius, is derived from the Greek word seirios, meaning "scorcher." Since Sirius is behind the Sun during the summer in the northern hemisphere, it was once thought that the combined light and heat of Sirius and the Sun was what led to the hottest part of the year, as well as a more ominous role of bringing evil and sickness to humanity. One of Sirius' nicknames, the "Dog Star", leads to the expression "dog days of summer." Some of the other names that Sirius has held came in part from ancient Hindu writing that refers to Sirius as "Tishtrya" (ruler of rain and water) and "Sukra" (the rain god). Other names include the Persian "Tir" (arrow), the Babylonian name "Kakkab-lik-ku" (star of the dog), the Assyrian "Kal-bu-sa mas" (dog of the sun), and the Akkadian "Mul-lik-ud" (dog star of the sun.)

The ancient Egyptians celebrated the pairing of Sirius and the sun, since it coincided with the annual flooding of the Nile. This revitalized Egyptian agriculture as well as Egyptian life in general. Sirius' "heliacal rising," or rising at dawn, thus marked the Egyptian new year.

Looking at Sirius through a telescope is by no means necessary since it is so dazzling just to the naked eye, but if you do so, you are treated to an impressive sight, reminiscent of looking toward the Sun as an interstellar voyager at the edge of the solar system. There is a companion to Sirius, called Sirius B (or "The Pup"). It's a white dwarf of magnitude 8.6. Periastron of this system occurred in 1994 and currently the separation is slowly increasing. Maximum separation of this system will be near the year 2020 when the two will be 11" apart, but Sirius B is always extremely difficult to detect due to the glare of Sirius A. Some speculation has been made of the behavior of Sirius B over the last two thousand years. Ancient writings by such notables as Ptolemy, Horace, and Homer describe the light of Sirius as ruddy, coppery, and redder than Mars. One theory suggests that Sirius B could have been in the red giant stage in recent times, which would have reddened the combined light of the system enough to garner such descriptions. Or it could have simply been a result of the Earth's atmosphere refracting the light of a bright star such as Sirius so that it appears to twinkle in a spectrum of colors when close to the horizon, and these writers took advantage of this for dramatic effect.

If you drop exactly 4 degrees south of Sirius, you will see the star cluster M41 (RA 6h 49.9m Dec -20° 44'). This is a 4.5 magnitude cluster easily seen with the naked eye in dark skies, and when spotted through a telescope reveals close to 100 stars of magnitude 7 and dimmer in a space about 1/2 degree across. The brightest star of the cluster is near the center and is a pinkish star of spectral type K3 II. The other bright members are giants of type K, G, and B.

NGC2354 is another cluster smaller and more subtle than M41 that has about 60 dim stars spread out in an area about 20' wide. Move about 1.5 degrees ENE of Wezen (Delta (δ) Canis Majoris), or center your scope at RA 7h 14.3m Dec -25° 44' for this 6.5 magnitude cluster.

A third cluster is the exquisite NGC2362, just 1.3 degrees to the northeast of NGC2354 at RA 7h 18.8m Dec -24° 57'. This is a tight cluster of about 40 stars 6' across crowded around the 4th magnitude star Tau (τ) Canis Majoris (also designated 30 Canis Majoris), a hot O9 type giant. The cluster roughly assumes the shape of an equilateral triangle with τ at the center. Studies indicate that this collection of stars is quite young, perhaps only 1 million years old.

Among the few galaxies that can be glimpsed in Canis Major is NGC2207. In terms of size (4.5' X 3') and brightness (mag. 12), this galaxy is typical of those which are found in Canis Major, but it shows a unique structure. It is tear drop shaped with the appearance of a double nucleus, which probably indicates that it is an interacting system of two galaxies. From Mirzam (Beta (β) Canis Majoris), move 3.5° to the SSW or center on RA 6h 16.4m Dec -21° 22.35'.

Canis Minor (kay' niss my' ner)
Abbreviation: CMi
The Little Dog
Best observed between January 15th and April 1st

The other, less conspicuous hunting dog of Orion is found south of Gemini and across the winter Milky Way from its companion, Canis Major. The bright star Procyon marks the heart of the constellation and forms the northeast point of an equilateral triangle with Orion's Betelgeuse and Canis Major's Sirius. Canis Minor, as its name suggests, is a small constellation. It encompasses only two bright stars and, even though it lies on the edge of the Milky Way, no deep sky objects.

Procyon's magnitude is a bright 0.4, making it the 8th brightest star in the sky. Its distance of 11.4 light years ranks it as number 13 of the nearest stars to Earth, and 5th nearest of the naked eye stars. Other vital statistics include a spectral type a bit hotter than the sun at F5, a luminosity of about six times the sun, and a diameter approaching 2 million miles, or over twice that of the sun. The name Procyon comes from the ancient Greeks. It translates "before the dog," since it rises mere minutes before the real dog star, Sirius. Because of its proximity, Procyon has a comparatively large proper motion, or movement relative to the sun. Its annual motion is 1.25" toward the southwest, which means that it will traverse a full degree in about 2,900 years.

Beta (β) Canis Minoris, or Gomeisa, is a hot B7 star 170 light years away. Gomeisa is a much brighter star than Procyon in total light output, but appears dimmer to us on Earth from its larger distance. If both were equally placed relative to the Earth at the standard distance of 10 parsecs, or 32.6 light years, Procyon would glow at magnitude 2.7, while Gomeisa would shine much more brightly at -1.1.

Similar to Sirius, Procyon has a dim white dwarf companion about 4.7" distant at a PA of 78 degrees. It was predicted to exist in 1861 due to the 40 year cycle of irregularities in the proper motion of Procyon, but because of the extreme differences in luminosity (Procyon B is 15,000 times fainter at magnitude 11), it was not visually observed until 35 years later through the 36-inch refractor at Lick Observatory. If you've been successful at observing the companion of Sirius, then you might want to try this one. However, Procyon B is a little more challenging, with over a full arc second less separation.

Two other double stars of Canis Minor share this characteristic of large luminosity differences. Eta (η) Canis Minoris at high power shows a magnitude 5.5 primary and a magnitude 11 secondary 4" apart. Nearby Gamma (γ) is a more comfortable 30" apart, but presents a more challenging secondary. It shines at magnitude 13 compared to the primary at magnitude 4.

Canis Minor is also home to some dramatic variable stars. All of them are classified as Long Period Variables, or LPV's. The three we will discuss are all in a west to east line in the northern half of the constellation. U (RA 7h 41m 20.7s Dec +08° 22' 50") goes from a moderate 8th magnitude to a practically invisible 13.8 over a period of 413 days. U is 44' northeast of a 7.1 magnitude star, which itself is 2.7 degrees due north of Procyon. Look 2.1 degrees due west of U (or hop 46' southeast from γ to a star of 7.1 magnitude, and then another 30' in exactly the same direction) for S Canis Minoris (RA 7h 32m 42.8s Dec +08° 19' 07"). Here we see a star fluctuating between magnitudes 6.6 and 13.2 during a 332 day period. On the extreme western edge of the constellation is V, a similar variable that goes from 7.4 to an invisible 15.1 over 366 days. This star is 5.2 degrees due west of γ (RA 7h 06m 58.8s Dec +08° 52' 41").

Capricornus (kap' rih korn' us)
Abbreviation: Cap
The Sea Goat
Best observed between July 15th and October 1st

The area of the sky where we find Capricornus is associated with the sea, since this is where we also encounter other constellations connected with water that are grouped together, such as Pisces, the fish; Eridanus, the river; Cetus, the whale or sea monster; and Aquarius, the water bearer. As far back as the Babylonians of 1000 BC, people have envisioned Capricornus as a creature which is mostly goat, but with its hindquarters replaced by the tail and fins of a fish. Myth tells us that this odd combination came into being after the monster Typhon surprised some gods picnicking beside the Nile. To escape the monster, the gods turned themselves into various animals to flee. But one of them, Pan, was indecisive. In his frightened state (we derive our English word "panic" from his name), he ran through the shallow water at the edge of the river. While he was waist deep, his impulsive outcome is our fish-goat hybrid.

Look south to find Capricornus. Its third magnitude and dimmer stars can be difficult to find. However, if you look in the area between Altair (the brightest star in Aquila on the eastern edge of the Milky Way), and Fomalhaut (the solitary first magnitude star sitting all by itself far to the southeast), you will see a shape in the stars resembling a fat letter V, or perhaps an orange segment.

The premier object in Capricornus is without doubt M30. Here is a fine globular cluster comparable to most other bright ones of its type that you may find elsewhere. It shines with a healthy magnitude 8 glow in a space about 10 arc minutes wide. Look to RA 21h 40m 25s Dec -23° 11' 00", or you can simply point to 5.2 magnitude 41 Capricorni located 3.5 degrees southeast of Zeta (ζ) Capricorni. M30 should then be in your low power field of view 1/3 of a degree to the northwest. Messier saw it in his telescope as a nebula, but soon afterward, other astronomers noticed that they could resolve it into stars and even described the cluster as being oval-shaped containing a spiral pattern of stars.

Looking opposite M30 to the western side of the constellation presents the spiral galaxy NGC6907. This is a suitable object for larger telescopes, since it is a faint 11.9 magnitude covering 3.3' x 2.7' of the sky. It sits at RA 20h 25m 6s Dec -24° 48' 30" in a blank area of the sky. You can star hop to it by starting at Psi (ψ) Capricorni. Move west 2 degrees to a 6th magnitude star, 1 degree west again to a pair of stars, one sixth magnitude and one seventh magnitude that are 8' apart, and then another 1.6 degrees west to NGC6907.

Cassiopeia (kass' ee oh pee' uh)
Abbreviation: Cas
The Queen
Best observed between October 1st and December 31st

Queen Cassiopeia plays a part in a complex story that also involves her daughter Andromeda as well as other mythological characters such as Perseus, Cetus, the gorgon Medusa, and Pegasus. She ruled an ancient land with her husband, Cepheus, until a character flaw resulted in her downfall. This was her boastfulness, a trait regarded with much disdain by the gods. To silence her own rants of how beautiful and talented she was, they sent punishment her way in the form of some tragic events. In the end, her country's coast had been destroyed by a sea monster (Cetus), and Cassiopeia herself was chained to a spot in the heavens to forever spin around the celestial pole as if to serve as a spectacle to all. Cassiopeia lies directly across the north celestial pole (marked by the star Polaris) from the Big Dipper, and her brightest stars form the shape of either the letter W or M depending on where she happens to be in your sky.

Despite Cassiopeia's ignoble status among the stars, her constellation lies along the Milky Way, and therefore it contains some wonderful deep sky treasures. A case could be argued for NGC7789 as being Cassiopeia's best star cluster. It doesn't contain any bright stars, but its richness is striking. It appears as a soft, cloudy area of magnitude 6.7 and 16' in size roughly halfway between Rho (ρ) and Sigma (σ)

Cassiopeiae or RA 23h 57m Dec +56° 44'. Through the telescope, however, it resolves into a spray of hundreds of points of light, each shining at magnitude 11 to 18.

The northwest corner of the constellation presents M52. It's comparable to NGC7789 by having over 100 stars of magnitude 9 to 13. You'll find M52 at 23h 24.2m Dec 61° 30', or follow the line formed by Alpha (α) and Beta (β) Cassiopeiae to the northwest a bit further than the distance separating those two stars. If we zoom to the eastern half of the constellation, we'll see M103, another star cluster located at RA 1h 33.2m Dec +60° 42'. This one is quite a bit leaner in the number of stars 1 degree ENE of 2.6 magnitude Delta (δ) Cassiopeiae. It measures about 7' in diameter with stars that appear to spiral outward from the center.

To find any galaxies here requires looking away from the obscuring band of the Milky Way. Moving your gaze southward, you may notice that the Andromeda galaxy (M31) lies in the same direction. Here is seen two far removed dwarf satellite galaxies of this great spiral. They are about 7 degrees to the north of M31, which corresponds to a true distance from M31 of close to 250,000 light years. If we start at Omicron (ο) Cassiopeiae and move one degree west, we find NGC185 at RA 0h 39m Dec +48° 20.3'. NGC147 lies one more degree west at RA 0h 33.2m Dec +48° 30.5'. Both galaxies are small elliptical systems very similar in size and shape: magnitude 10.1 and size of 12' x 10' for NGC185; and magnitude 10.4, size 13' x 8' for NGC147. To observe either galaxy requires a fair amount of aperture (such as 8") and a dark sky. You will see just a faint brightening of the background sky.

If you are eager for another challenge, then NGC281 is appealing. This is a faint emission nebula 1.7 degrees east of magnitude 2.2 Alpha (α) Cassiopeiae (RA 0h 52.8m Dec +56° 37'). A nebula filter improves your chances of seeing this diffuse cloud of gas of magnitude 7 and spread out over 35' of the sky.

One of the best double stars in Cassiopeia (or anywhere) is Eta (η) Cassiopeiae. Shining at magnitudes 3.5 and 7.5, they are near maximum separation, currently about 14". Their colors are contrasting white and gold.

Centaurus (sen tor' us)
Abbreviation: Cen
The Centaur
Best observed between April 1st and May 15th

There are two centaurs (half man, half horse) in the sky, Sagittarius and Centaurus. The former gets most of the attention because it sits at a much higher declination and is immersed in the best part of the Milky Way that is visible from the Northern Hemisphere. Centaurus, on the other hand, is at least half obscured by the horizon from most locations in the Northern Hemisphere, and it has only a few interesting objects.

From mythology we get the story of Chiron, the smartest and wisest of the centaurs. He was a skillful artist, hunter, and physician and passed some of his great knowledge to mankind, including his knowledge of the sky. Chiron met his demise when he was accidentally wounded by one of Heracles' poisonous arrows. Since Chiron was immortal, he didn't die, but was in continuous pain. He pleaded with the gods to release him from his misery and allow him to die, finally having to offer his own life for the release of Prometheus, who had been imprisoned for the crime of stealing fire from the gods and giving it to mortal man. Zeus placed Chiron in the sky, but since the northern sky was already filled with constellations, the only place for Chiron was far in the south. We have to wait until Centaurus is on the meridian before we can see most of the constellation. Look below Spica, the bright, blue-white star of Virgo, and below the tail of Hydra. The upper body of Centaurus is formed from second and third magnitude stars scattered low in the sky and is centered on the meridian near 11pm on May 1st.

One of the unique aspects of Centaurus is that it is one of only two constellations having two stars that are brighter than 1st magnitude (Orion is the other). The problem is that these two stars (Alpha (α) and Beta (β) Centauri) are only visible if you are south of 29 degrees north latitude. α Centauri, of course, is noted for being the closest star system to Earth (4.3 light years), and the fact that it is a strikingly bright double star that is slowly widening from its current separation of about 3 arc seconds.

Most of Centaurus's deep sky wonders are below the horizon for those of us confined to high northern latitudes, but there are a few magnificent ones that rise high enough to view. Near the northern edge of Centaurus is NGC5253, an 11.1 magnitude galaxy of an oval shape spanning 5' x 1.9'. It shows little structure, but a gradual increase of brightness toward the center. It's a little hard to find in an empty area of sky a full 6.5 degrees northeast of Iota (ι) Centauri, or RA 13h 39m 56s Dec -31° 38' 41".

Going back to ι Centauri, we'll see a galaxy, NGC5102, just 17 arc minutes east northeast of this 2.7 magnitude star. It measures a sizable 8.8' x 2.9' and shows a bright core contrasting with the fuzzy edges. Zoom in to RA 13h 21m 58s Dec -36° 37' 47" to see this magnitude 10 galaxy.

Barely above the horizon is NGC4945, a wonderful galaxy of magnitude 9.2 and a large size of 19' x 3.8'. Through the eyepiece, which will be filled from edge to edge at moderately high power, it will appear as a soft spear of light with strong mottling to one side through large instruments. Look to RA 13h 5m 24s Dec -49° 29' 06", or 4 degrees east of Gamma (γ) Centauri between Xi1(ξ1) and ξ2 Centauri.

Rounding out our collection of bright galaxies is a favorite of galaxy hunters, NGC5128, or Centaurus A. Using ι Centauri again as our starting point, slide your scope south 6.4 degrees. This galaxy is famous for being a bright radio source, presumably from an active black hole at its center. In a telescope, NGC5128 appears as a very large, round galaxy occupying 25' x 20' and shining brightly at magnitude 7.2. Most telescopes will show its most prominent feature, a vast lane of dust crossing its center. Large instruments in the range of 12 to 14 inches and larger will reveal that the dust band is made of two thick strands with intervening dust mottling the middle. Its location is RA 13h 25m 29s Dec -43° 1' 00".

Besides galaxies, Centaurus is home to two bright, large globular clusters. Let's start with the lesser of the two, NGC5286. Here we have a magnitude 7.6 cluster measuring 9.1' in diameter that is very easy to find. (For comparison, the well-known Hercules cluster, M13, is of magnitude 6 and a size of 16'.) Start at magnitude 2.3 Epsilon (ε) Centauri and move 2.3 degrees northeast to the next brightest star (M Centauri) that you will see at magnitude 4.6. NGC5286 is only 4 arc minutes northwest of this star (RA 13h 46m 24s Dec -51° 22' 00"). NGC5286 will be impressive in any scope.

Now we come to the most impressive globular cluster in the heavens, NGC5139. Only 47 Tucanae in the far southern sky can compare with it. So bright is this cluster that it's easily visible with the naked eye, and before it was proven to be a cluster with telescopic observations, it was given the Bayer designation of Omega (ο) Centauri as if it were a star. Spanning a huge area of 36' (bigger than the full moon), and shining at magnitude 3.7, Omega Centauri evokes various interjections from the first time or occasional observer. Look just under 5 degrees west of 2.5 magnitude Zeta (ζ) Centauri or RA 13h 26m 48s Dec -47° 29' 00". It doesn't take a scope of much size to begin to resolve the stars in Omega Centauri, and if you are working with large aperture, the swarm of a million stars inside this cluster is truly dizzying. No doubt those with strong imaginations will see patterns, shapes, and streamers within the cluster, including two areas of lower star densities that seem to form darker patches resembling two dull eyes looking back at the observer.

It is true that many of the objects in Centaurus suffer a bit from their low declinations. Observers closer to the equator are more fortunate to have the constellation higher in the sky. But don't hesitate to take advantage of some clear spring skies and seek out the wonders of Centaurus that are nevertheless visible in our skies.

Cepheus (see' fee us)
Abbreviation: Cep
The King
Best observed between September 1 and November 1st

This constellation crosses the meridian at 9pm during the middle of October and can be found in the northern sky between the cross-shaped constellation of Cygnus, and the "W" pattern of Cassiopeia. Its boundaries stretch all the way to Polaris (the north star), so since it is situated so near the pole, most of the constellation remains above the horizon at all times. Even its most southerly stars rise again no more just a few hours after setting.

Cepheus was king of an ancient land in northern Africa, with Cassiopeia as his queen. These two, in addition to their daughter Andromeda, form a family portrait of sorts in the sky, all sitting side by side in the area of the heavens north of +30 degrees declination and between 21h and 2h RA.

In 1784, John Goodricke discovered that Delta (δ) Cephei, in the extreme southern part of the constellation, exhibited variations in brightness that modern studies have precisely measured to be 5.366341 days. The variability results from an actual pulsation of the star rather than other phenomena like eclipses or outbursts. A multitude of these pulsating stars in the Milky Way and other galaxies have since been discovered, and are all called Cepheid variables after this star. This stellar class is very important to astronomers, since the period of pulsation (which generally ranges from a few hours for the dimmest Cepheids to as long as 50 days for the brightest ones), correlates to the star's luminosity. The brighter the Cepheid, the longer the period. Thus, Cepheid variables can be used as accurate measuring sticks for determining distances to other galaxies that also have this type of star when the period and apparent magnitude of these stars are known.

Although δ is an easier star to see, its magnitude range of 3.6 to 4.3 is less noticeable than U Cephei, having a range from magnitude 6.8

to 9.2. U is an eclipsing binary at RA 1h 2.3m Dec +81° 52.5', or if you look in the area one-third the distance from Gamma (γ) Cephei to Polaris and move 1h 25m east of that line (a short distance at this declination, about 1.5 degrees), you will see a group of four stars forming a crooked line 23' long and all near 8th magnitude. U is the southernmost of this group and, if glowing at its brightest, will outshine the other three. If at its dimmest, it will fall short of the others. Its period is about 2.5 days and takes only 8 hours to dim and return to normal after spending 2 of those hours in its eclipsed phase.

Now, point your scope to Mu (μ) Cephei. This is the famous "Garnet Star," which gets its nickname from the rich orange-red color it shows. This is an irregular variable with fluctuations spanning hundreds of days, but it always remains visible to the naked eye under suitable conditions. It forms a wide triangle with Alpha (α) and Zeta (ζ) Cephei, and if you have the patience, you may notice its variability from magnitude 3.7 to 5.0. As a point of interest, μ is the north polar star of the planet Mars.

Among the brighter deep sky objects in Cepheus are NGC6939 and NGC6946. NGC6939 is an open star cluster made of a scattering of dozens of stars ranging between 12th and 15th magnitude in an area about 8' in size. NGC6946 is a magnitude 11 spiral galaxy of about the same size to the southeast. It looks slightly oval and is generally of uniform brightness across its face. I mention these together because they are only about 40' apart and present a contrasting duo in an eyepiece capable of about a 1° field of view. Center your scope at RA 20h 33m Dec +60° 20' (or about 2.5 degrees south of Theta (θ) Cephei and slightly east) and you'll see this pair.

Sharp eyed observers will see the faint nebulosity of NGC7023. It surrounds a 7th magnitude star at RA 21h 0.5m Dec +68° 10', or 3.5 degrees southwest of Beta (β) Cephei. Its central glow is roughly rectangular in shape and fades quickly to the edge of its total area of 18'.

There are many more galaxies, nebulae, and star clusters in Cepheus. You may use a star chart as a guide, or simply browse the constellation on your own journey of discovery.

Cetus (see' tus)
Abbreviation: Cet
The Whale, or Sea Monster
Best observed between October 15th and December 31st

Not too far south of the ecliptic and neighboring Taurus to the west is the constellation of Cetus. Through the ages, Cetus has been seen as several incarnations of sea monsters. The ancient Mesopotamians saw in the stars of Cetus a cosmic dragon named Tiamat. Traditional mythology depicts Cetus as the sea monster who was to receive Andromeda, daughter of Cepheus and Cassiopeia, as a sacrifice to act as retribution for her mother's boastfulness. Perseus, however, saved her by destroying Cetus with the head of the Gorgon called Medusa. All of these characters, with the exception of Medusa and Cetus itself, are marked by constellations in the same general area of the northern sky. Medusa has no constellation of her own, and Cetus lies farther south in the "wet" part of the heavens. Here is has the company of other constellations with connections to water, like Aquarius the water bearer, Eridanus the river, and Pisces the fish. Today, Cetus is often illustrated as a more earthly and docile sea creature: a whale.

Cetus is the location of Mira (or Omicron (ο) Ceti), the first variable star to be discovered. The German astronomer, David Fabricius, was the first to note Mira's variability in 1596. Since that time, it has taken dedicated observations to discover that Mira has a period of 331.65 days, and its magnitude varies dramatically from a fairly bright 2.0 to a feeble 10.1. Among the many categories of variable stars, Mira belongs to the type called long-period variables (LPV's), which are often called Mira stars. Interestingly, the Latin root "mira" means "wonderful," and it is the basis for the English word "miracle." Indeed, based on Mira's behavior, it may have seemed like a miracle to some people to learn that the universe contained such an enigma and was not the unchanging realm it was once thought to be.

Since Cetus lies far from the galactic equator, it is free from the obscuring dust that effectively hides distant galaxies from view. Therefore, Cetus is brimming with them and is a galaxy hunter's paradise. However, most are small and dim (a trait common to many galaxies), and will require some determination to locate. The easiest to find is M77 (RA 2h 42m 40s, Dec -0° 01'), a 10th magnitude spiral galaxy turned mostly face-on to us not quite one degree to the southeast of Delta (δ) Ceti. You may very well see others if you look within two degrees north and east of M77, although it may prove difficult seeing these 11th and 12th magnitude objects. The same holds true for an area 8.5 degrees due south of M77, but if they show themselves, you'll be treated to a cluster of four galaxies, all about 30 arc minutes or less apart. Spirals, barred spirals, and elliptical galaxies are scattered all across Cetus. A typical star atlas may show more than thirty of them. But remember, they are all small with magnitudes dimmer than 11, so get your telescope with the best aperture and look closely for them.

For a change of pace, look 6 degrees north and slightly east of Beta (β) Ceti, the constellation's brightest star. There you'll find NGC246, a planetary nebula of about 9th magnitude (RA 0h 47m 05s Dec -11° 51'). Large telescopes may show a mottled appearance, otherwise it will look like a fuzzy star about 4' in diameter. High magnification will darken its surroundings and make it easier to study

Gamma (γ) Ceti is a double star by the name of Kaffaljidhma, an Arabic term for "the sea monster's head." This name previously referred to all the stars forming the circle marking Cetus' head, but now only identifies this single star. Its components are of magnitudes 3.7 and 6.4 with a PA of 295 degrees. High magnification is necessary to overcome the small separation of 3 arc seconds of this fine yellow/blue pair.

Columba (koh lum' buh)
Abbreviation: Col
The Dove
Best observed between January 1st and February 28th

This obscure little constellation largely escapes notice by many stargazers because of two circumstances. First, it is in a part of the sky containing many other bright constellations that present luring deep sky objects to steal attention away from it. Secondly, its southern location means that it is in its most favorable spot above the horizon for a comparatively short period of time before sinking back into the murky atmosphere. Columba made its first appearance in star charts published around 1679, having been split off into its own constellation from stars that were originally included as part of Canis Major. It is considered to be the dove mentioned in the book of Genesis that Noah released from the Ark after the rains had subsided. The dove returned to the Ark with an olive branch, signifying not only that flood waters had receded to uncover pieces of land, but also the fact that God had made his peace again with His children on Earth.

To find Columba, look directly below the well-known constellation of Orion. Your eyes will pass through the dim stars of Lepus, and eventually you will see an irregular zigzag of 3rd and 4th magnitude stars. Columba is only slightly smaller in area compared with neighboring Canis Major and Lepus. In the middle of January, Columba is rising at 5pm, is at its highest by 9:30pm, and is well on its way to beginning its descent below the western horizon by midnight.

An interesting (but visually mundane) star is in the northern half of Columba. This is Mu (μ) Columbae. It is one of three stars (the others being AE Aurigae and 53 Arietis) called "runaway stars" because of their high velocity through space. The three are moving in opposite directions from each other but are thought to have originated in the same nebulous regions of Orion. μ is moving through space relative to the sun at close to 120 kilometers per second, which means it was launched on its present course between 2 and 3 million years ago. Perhaps μ and its counterparts were members of close binary pairs, and their companions exploded as supernovae to release the stars from their grip, changing the orbital velocity to a straight line velocity. By simply looking at μ Columbae, it is difficult to imagine its incredible speed.

The best deep sky objects are all located in the southwest section of the constellation. NGC1808 is a bright spiral galaxy of magnitude 10.8 at RA 5h 7m 43s Dec -37° 30' 51", or 1.9 degrees south southeast of Gamma (γ) 2 Columbae. Its bright center spans about 4' x 1' but, including its faint outer arms, increases its full expanse to about 6.5' x 3.9'. This barred spiral galaxy is near 40 million light years from Earth and about 35,000 light years across. There are a couple of unique features of this galaxy. Its disk is fairly warped and it has such a tremendous rate of star formation near its center that it can be put in the category of "star burst" galaxies. It is possible that its structure and behavior can be linked to its proximity to nearby galaxy NGC1792 described below.

NGC1792 is a similar spiral galaxy, presenting a magnitude of 10.7 and an expanse of 5.2' x 2.6'. It is only 40' to the southwest of NGC1808 or RA 5h 5m 15s Dec -37° 58' 47". It can be spotted in scopes as small as 3" and shows mottling in large scopes of 12" or larger.

Move 2.6 degrees southeast of NGC1792 (RA 5h 14m 6s Dec -40° 3' 00") to find a very good globular cluster. This is a rich, bright cluster, about 54,000 light years distant, and well worth the time to hunt down. Its thousands of stars are packed into a dense cluster with a bright center. The overall magnitude is 7.3 and a size of 11'.

Coma Berenices (koh' muh bair en ees' eez)
Abbreviation: Com
Berenice's Hair
Best observed between March 15th and May 31st

Stargazers are presented with the best collection of galaxies the sky has to offer in the springtime skies. Here is where we find the Virgo-Coma cluster of galaxies. Most of these galaxies are located in the constellation of Virgo. However, a large number are within the dim constellation of Coma Berenices, which borders Virgo to the north.

The name of this constellation comes from Berenice II, who was the queen to Ptolemy III, ruler of Egypt in about the year 240 BC. When Ptolemy III left to fight an especially dangerous battle, Berenice II pledged to sacrifice her long, golden hair (of which she was very proud) if her husband returned victorious. When he did, she placed the hair in a temple honoring the goddess Aphrodite. To their dismay, the hair disappeared from the temple a short time later, so the court astronomer, Conon, comforted the royal couple by saying that the goddess was so enamored by the gift that she placed the queen's hair in the heavens so that everyone could admire it. You can find Berenice's hair by looking between the nearby constellations of Leo bordering to the west, Boötes to the east, and Virgo to the south.

To the naked eye, the most distinguishing feature of the constellation is the large open cluster Melotte (or Mel) 111. This appears as a hazy spot sprinkled with a small handful of brighter, naked eye stars near the northwest section of Coma Berenices. This is one of the nearest star clusters to Earth at 260 light years, and for that reason, is spread out across a large chunk of the sky. Binoculars are best for viewing this cluster since it measures a full 5 degrees across. The most prominent shape of the cluster reminds me of a vintage propeller-driven fighter plane with its wings folded up for storage on an aircraft carrier.

Overall, there are about 3,000 galaxies in the Virgo-Coma cluster. While over 200 bright members can be found within the boundaries of Virgo, more than 60 similar objects can be found in Coma Berenices. Scanning the area will bring galaxy after galaxy into view. M64 deserves special mention, since its bright 8.0 magnitude glow is easy to find slightly less than one degree to the northeast of the star 35 Comae Berenices (RA 12h 56.7m Dec +21° 41'). This galaxy is nicknamed the "Black Eye Galaxy" because of a huge dark mass of dust wrapping around the bright nucleus. At a dark, clear site you should see close to half of the inner disk darkened by the dust cloud, in contrast to the outer glow of the galaxy's oval.

NGC4565 is a favorite to those familiar with it. It is turned edge-on to our line of sight, and measures 15' in its longest dimension, and only 1' in width. It appears much like a knife blade against the darkness of space, and running along its length is an easily noticeable dust lane which bisects the egg-shaped nucleus. Find NGC4565 7' in RA east of 17 Comae Berenices, the closest 5th magnitude star of Mel 111, or RA 12h 36.3m Dec +26° 00'.

We can summarize other Messier galaxies as follows:

M85	Mag. 10.5 Spiral	RA 12h 25.4m Dec +18° 11'
M88	Mag. 10.9 Spiral	RA 12h 32.0m Dec +14° 25'
M91	Mag. 11.1 Spiral	RA 12h 35.5m Dec +14° 30'
M98	Mag. 11.4 Spiral	RA 12h 13.8m Dec +14° 54'
M99	Mag. 10.5 Spiral	RA 12h 18.8m Dec +14° 25'
M100	Mag. 10.8 Spiral	RA 12h 22.9m Dec +15° 49'

Closer to our home galaxy are two bright globular clusters: M53 and NGC4147. M53, magnitude 8 and 16' in diameter, has a broad nucleus saturated with stars 1 degree northeast of Alpha (α) Comae Berenice. NGC4147 is one-third the size of M53 and has few, if any, resolvable stars, but its obvious nucleus is easy to find 6.5° northeast of Beta (β) Leonis.

Corona Australis (kor oh' nuh oss tray' liss)
Abbreviation: CrA
The Southern Crown
Best observed between June 15th and August 15th

Corona Australis is one of the original 48 constellations envisioned by the second century astronomer Ptolemy. It is a half circle of 4th magnitude and dimmer stars under the teapot asterism of Sagittarius, and east of the lower curve of the tail of Scorpius. Most depictions of Corona Australis show it as a laurel wreath. It is considered to be the crown of Chiron (the centaur) or a duplicate of Corona Borealis (the constellation known as the northern crown that belonged to Ariadne, the daughter of the king of Crete, Minos.).

Corona Australis lies near the heart of the summer Milky Way, so the deep sky objects that we expect to see are star clusters and nebulae. We can start with NGC6541, a bright globular cluster in the southwest corner of the constellation. One way to find this bright, 6.6 magnitude cluster is to star hop from some bright stars in the tail of neighboring constellation, Scorpius. Start with Lambda (λ) Scorpii (the tip of Scorpius's tail) and move southeast through the tail stars of Kappa (κ) and Iota (ι). Keep going in this same direction two and a half degrees to a 4.9 magnitude star, then repeat this step two and a half degrees to another 4.9 magnitude star. NGC6541 is then less than half of a degree southeast of this last star. You can also use your setting circles or computerized telescope and dial up RA 18h 8m Dec -43° 42'. Here is a fine globular cluster which you may or may not be able to resolve depending on the size of your telescope and atmospheric conditions. Its overall magnitude of 6.6 means it is easy to spot, and the size of 13' makes a good view at moderate power.

Nearby at 1.7 degrees west-southwest of NGC6541 (or RA 17h 59m Dec -44° 16') is another globular cluster, NGC6496. It straddles the boundary between Corona Australis and Scorpius. Here we have a more subdued cluster glowing at magnitude 9.2 and having about half the size of NGC6541, but it is still well worth a look. This cluster appears as a loose ball of stars.

For experienced observers, there is an elusive planetary nebula at the opposite side of the constellation at RA 19h 17m 24s, Dec -39° 37' 00", or 1.5 degrees east of Beta (β) Coronae Australis. This is IC1297, a small nebula measuring a mere 0.1 arc minute across. At high power, you should be able to distinguish this as a diffuse circle, otherwise it will be difficult to see it as anything different than the surrounding stars.

A diffuse area of reflection and dark nebulosity is found in the northeastern section of Corona Australis. Here we have a collection of nebular bodies composed of NGC6729, NGC6726, and NGC6727. Although it sounds complex, they occupy a fairly small area of sky measuring about 5 arc minutes. One of these nebulae, NGC6729, is illuminated by the variable star R Coronae Australis, so the responding nebula also varies. This star is an irregular variable which fluctuates between magnitude 9.7 and 12, and lights up the small, comet shaped nebula. So active is this star that it has been observed changing by as much as 2 magnitudes in only a few days time. NGC6726 and NGC6727 are side by side, more or less in contact with each other, and forming a double lobed structure of about 2 arc minutes in the longer dimension. Find this group by centering your scope near 19h 1m 50s Dec -36° 55', or just less than 1 degree west of magnitude 4.2 Gamma (γ) Coronae Australis.

After finding some of these elusive objects, we can relax a little and go after an easy double star. It's one of the brighter stars in this constellation but, nevertheless, is just within naked eye limit in dark skies. The star is Lambda (λ) Coronae Australis, sitting just to the west of the brightest arc of stars forming the most apparent structure of the constellation. λ is a pair of magnitude 5 and magnitude 9 stars separated by a wide 29 arc seconds, so it can be separated by the smallest of telescopes at very low power.

Often we forget about Corona Australis because it is adjacent to the wonderful offerings of Sagittarius and Scorpius, but sliding your scope a little more to the south will bring you into its environs which offer some wonders of its own.

Corona Borealis (kor oh' nuh bor ee al' iss)
Abbreviation: CrB
The Northern Crown
Best observed between April 15 and June 30th

This constellation, with its beautiful name and lovely shape, is one of the smaller ones in the sky. It sits between Boötes, with its bright star Arcturus, and Hercules, reaching the meridian just after 10pm on June 15 and between 1am and 2am in late April. With its 2nd, 3rd, and 4th magnitude stars, it assumes the shape of an incomplete circle.

The brightest stars of Corona Borealis form an isolated group and are therefore easily recognizable. The most ancient story behind the crown involves the order by the king of Crete, Minos, that the city of Athens send him 14 young men and women each year to be sacrificed to the Minotaur, (a creature that was half man, half bull) living in the inescapable Labyrinth. One year, a man by the name of Theseus (the son of Athens' own king) was sent as one of the sacrifices, but Minos' daughter, Ariadne, fell in love with him. She gave him a sword and a ball of string before he was sent into the Labyrinth. He successfully killed the Minotaur with the sword and found his way out of the Labyrinth by following the string that he had trailed behind him. Theseus and Ariadne left Crete for Athens, but Theseus had a change of heart and decided to leave her during the journey. As Ariadne wept, the god Dionysus found her and presented himself to proclaim his love for her, but Ariadne was distrustful. To prove he was a god, he took the crown he was wearing and threw it into the sky, where it turned into the stars of Corona Borealis.

In North America, the Shawnee tribe of Native Americans saw this constellation as a circle of dancing women appearing as stars in the heavens. The circle is incomplete because one of them decided to leave for Earth to live with a mortal.

You won't find any nebulae or galaxies in Corona Borealis, but what you will encounter are some very intriguing stellar objects. Epsilon (ε) Coronae Borealis is a very challenging double star along the east side of the crown. The primary star shines at magnitude 4.1, which is easy enough to find. Spotting the companion will require some skill, however. First of all, it sits only 2 arc seconds from the primary, and secondly, it glows reluctantly at magnitude 12. You will see that the secondary is lost in the light of the primary unless very high magnification is used to separate them.

Two much easier binaries are Sigma (σ) and Zeta (ζ) Coronae Borealis. ζ is a pair made of one 5th and one 6th magnitude stars separated by 6.3 arc seconds. The two stars are identical in color, both being of spectral type B. σ is virtually an exact copy of ζ. Except for their slightly different PA's, you'll have trouble telling the two apart.

The variable star R Coronae Borealis is the epitome of what is called an irregular variable. Since it was first observed in 1795, the star has followed no predictable pattern. Maximum brightness is near 6th magnitude, and it can hold steady at this brightness for either several years, or as little as a few days. And at other times, it fluctuates wildly between 6th and 15th magnitude. Observations have shown these fluctuations can last anywhere from one to ten years before stabilizing back to maximum again. It is anyone's guess how bright it may be at any date in the future.

Earlier, I mentioned that no galaxies are apparent in Corona Borealis. There is, however, a galaxy cluster that rivals the well known Virgo/Coma cluster in terms of numbers and actual size in space. The difference is that this cluster of over 400 galaxies is so much more distant that, from Earth, its brightest members are fainter than 16th magnitude. Unless you have a 45-inch telescope, let your imagination ponder this vast collection of spirals and ellipticals.

Corvus (kor' vuss)
Abbreviation: Crv
The Crow, or Raven
Best observed between March 1 and April 30th

Zoologists know the term "Corvus" as the genus to which crows belong. Astronomers are also familiar with the term, since a constellation by the same name makes its appearance in the mid-evening hours during the spring months, reaching the meridian at 11pm during the middle of April. If you look southwest of Spica in Virgo, you will see an irregular quadrilateral of stars that is the constellation of Corvus. It is one of the smaller constellations, measuring 184 square degrees (Orion, by comparison, occupies 594 square degrees.)

One legend states that the god Apollo gave Corvus a cup to fetch some spring water. Along the way, the crow was tempted by a fig hanging from a tree, but since the fruit was still green, the crow couldn't resist waiting for it to ripen. When he returned, Corvus explained his tardiness by bringing back a water snake along with the full cup of water, saying the snake had attacked him. Apollo was too wise to believe the foolish bird and banished Corvus, the cup (Crater), and the water snake (Hydra) to the sky. Another legend says that Apollo was suspicious that Coronis, with whom he was having an affair, was being unfaithful, so Apollo sent Corvus, who then was a silver bird, to spy on her. Corvus reported back that Apollo's suspicions were correct, so the god slew Coronis and sent Corvus to Hades where his feathers were burned black.

The four bright stars of Corvus, sitting southwest of Spica (the brightest star in the constellation Virgo), was seen as a tent by the Arabs. In fact, Alpha (α) Corvi is named Alchiba, the Arabic word for "tent." This is a relatively nearby star lying 68 light years from Earth. Delta (δ) Corvi, named Algorab (of Arabic origin meaning "raven's wing"), is a pleasant double star featuring a primary of magnitude 3.1 and spectral type of B9 V, and a dimmer secondary shining at magnitude 8.4 and having a spectral type of K2 III. The two are separated by a comfortable 24.2 arc seconds, so splitting them in a telescope is an easy task. You could even consider δ forming a visual double with nearby Eta (η) Corvi, 37 arc minutes away and of magnitude 4.3.

The most intriguing deep sky object in Corvus undoubtedly has to be NGC4038, also known as the "Ringtail Galaxy". The coordinates are RA 12h 01.8m Dec -18° 51.3', or about 14 minutes in RA west and 1.2 degrees south in Dec from Gamma (γ) Corvi. This is a magnitude 11 galaxy measuring 2.5 arc minutes on each side. Through medium-sized telescopes, this galaxy exhibits a heart-shaped structure, resulting from what is apparently a collision between two separate galaxies. Although too dim to see through most backyard telescopes, two long filaments extend tens of thousands of light years on each side of the main body of the galaxy. It has been determined that these long trails of stars are the result of gravitational interaction from the combined mass of the two systems. A Hubble Telescope image showed vast regions within the two halves of the merging pair, where stars are forming at an incredible rate because of the forces involved with this process.

A short hop of 0.7 degrees to the southwest will reveal another galaxy with a similar brightness of magnitude 11.6 and a size of 2.0 x 1.7 arc minutes. This is NGC4027 (RA 11h 59.5m Dec -19° 15.3'). At an actual distance of only about a million light years from NGC4038, it appears to be a third member of this small group. In fact, it is an irregularly shaped spiral that seems to be distorted by the gravity of its interacting neighbor. Through a telescope, it shows little overall detail other than a smooth oval.

A fairly easy planetary nebula floats within the heart of Corvus at RA 12h 24.6m Dec -18° 43.3'. With the designation of NGC4361, this 10.5 magnitude planetary can also be found by moving 5 minutes west in RA and 2.3 degrees south of Delta (δ) Corvi. The diameter of this nebula is about 80 arc seconds (or almost exactly the same size as the often observed M57 in Lyra) and, depending on the size of your telescope, gives you a decent opportunity at glimpsing its central star shining at magnitude 13. The nebula is almost perfectly round with a hint of an oval shape, and its surface is smooth with no apparent mottling.

If you can tear yourself away from the galaxy clusters of Virgo and Coma Berenices, look southward to Corvus, where you can find more individual treasures.

Crater (kray' ter)
Abbreviation: Crt
The Cup
Best observed between March 1st and April 30th

Crater plays a role in an elaborate story involving the nearby constellations of Corvus (the crow), and Hydra (the water serpent). The myth tells of Apollo, who sent Corvus for a cup (represented by Crater) of spring water. Beside the spring, Corvus spotted a fig ripening on a nearby bush. The fig looked so delicious that the crow waited for it to ripen. When he returned to Apollo, he brought along with him a water serpent (Hydra), saying that the serpent had attacked him, and that was the reason it took him so long to return. Apollo saw through the lie and banished all three to the sky. To find Crater, put yourself under some dark skies, since Crater's brightest star is magnitude 3.5 and most others are dimmer than magnitude 4. Look either 30 degrees (or about three times the width of your fist held at arm's length) west of Spica, the brightest star in Virgo, or 30 degrees south of the hindquarters of Leo. With not too much imagination, you should be able to see the shape of a cup tilting to the east.

Looking toward Crater, which crosses the meridian just before midnight during the ides of March, will give you a view about 25 degrees away from the southern Milky Way. Through the sprinkling of nearby stars lies an unobstructed view toward faraway galaxies. Most of these are dim and challenging, but a few are within the grasp of small telescopes.

On our way to these galaxies, let's stop at double star Gamma (γ) Crateris. This pair, 78 light years away, is separated by 5 arc seconds, allowing it to be easily split at high power. The primary shines brightly at magnitude 4.1. The secondary is several times dimmer at magnitude 9.5. Both show a pure white color.

NGC3511 is one of the brightest galaxies in Crater. It is located 2 degrees west of Beta (β) Crateris. While it shines at magnitude 11.6, its light is spread over a broad oval measuring 5.7' x 2.0'. Nearby (only 10 arc minutes to the southeast) is NGC3513. Here is an "S"-shaped barred spiral, glowing faintly at magnitude 12.1 and measuring 2.8' x 2.3'. If you center your scope on RA 11h 3m 35s Dec -23° 9' 57", you will have both galaxies in a low power eyepiece.

Within or near the cup of Crater are three more galaxies of note. NGC3887 is a spiral located one and a half degrees north northeast of Zeta (ζ) Crateris or RA 11h 47m 5s Dec -16° 52' 22". This is a tight spiral of magnitude 11.6 and is 3.3' x 2.5' in expanse. This galaxy shows a bright center with tightly wrapped arms that quickly disappear beyond its fuzzy boundaries.

Just above the lip of the cup is NGC3962. You'll find this elliptical galaxy about a third of the way between Eta (η) and Theta (θ) Crateris and 1.5 degrees above the line connecting these two stars. Or you can dial up RA 11h 54m 40s Dec -13° 58' 23". NGC3962 shines at magnitude 11.9 but perhaps is easier to see because of its bright, condensed core. The entire galaxy occupies 2.5' x 2.2'.

Just over a degree north of Epsilon (ε) Crateris is NGC3672, yet another galaxy glowing with a fairly bright magnitude of 11.7. This spiral galaxy, centered at RA 11h 25m 13s Dec -9° 47' 43", is a spiral with a fairly bright center and mottled spiral arms.

The galaxies thus far described only scratch the surface of the total number of galaxies present in Crater. However, most of them are reluctant to show themselves in small backyard telescopes. Those observers with larger instruments may want to search for some of these others listed below.

NGC3571: magnitude 13, RA 11h 11m 30s Dec -19° 18' 04".
Size: 2.5' x 0.7'. Bright nucleus, distorted arms.

NGC3635: magnitude 12.7, RA 11h 20m 31s Dec -9° 0' 49".
Size: 1.2' x 0.9'. A very small and faint spiral.

NGC3637: magnitude 12.9, RA 11h 20m 40s Dec -10° 16' 32".
Size: 1.6' x 1.5'. Paired with NGC3636. 3' from 6.5 magnitude star.

NGC3715: magnitude 12.0, RA 11h 31m 32s Dec -14° 13' 57".
Size: 1.4' x 0.9'. Very small, round spiral.

NGC3732: magnitude 13.0, RA 11h 34m 12s Dec -9° 51' 00".
Size: 0.6' x 0.6'. Small, bright center.

NGC3865: magnitude 13.0, RA 11h 44m 52s Dec -9° 13' 59".
Size: 2.0' x 1.5'. Diffuse spiral.

NGC3892: magnitude 12.0, RA 11h 48m 0s Dec -10° 58' 00".
Size: 2.0' x 1.3'. Round spiral.

NGC3955: magnitude 12.6, RA 11h 53m 58s Dec -23° 9' 54".
Size: 2.0' x 0.9'. Spiral tilted almost edge-on.

NGC3956: magnitude 12.9, RA 11h 54m 1s Dec -20° 34' 00".
Size: 3.4' x 1.0'. Edge-on spiral.

NGC3957: magnitude 12.9, RA 11h 54m 1s Dec -19° 34' 06".
Size: 3.0' x 0.7'. Lenticular galaxy.

NGC3981: magnitude 11.9, RA 11h 56m 7.2s Dec -19° 53' 42".
Size: 5.2' x 2.3'. Double-armed spiral.

IC2627: magnitude 12.6, RA 11h 9m 53s Dec -23° 43' 35".
Size: 2.7' x 2.3'. Face-on spiral. Star like nucleus.

Cygnus (sig' nuss)
Abbreviation: Cyg
The Swan
Best observed between July 1st and September 30th

There are many stories behind this constellation. One describes Cygnus as the Greek god Zeus after he disguised himself as a swan to seduce Leda, wife of a Spartan king. Another tells of Orpheus, a musician who was changed into a swan and lofted into the sky next to his harp (the constellation Lyra). And then there is the dramatic story of a young man called Cygnus who was the friend or brother of Phaethon, the son of Apollo. Phaethon borrowed the Sun chariot for a quick ride, but soon lost control of the vehicle. To stop the chariot, Zeus was forced to kill Phaethon, who fell into the river Eridanus. Cygnus frantically swam through the waters in an attempt to pull Phaethon back to shore for burial. Zeus was touched by the devotion Cygnus showed to his friend, so he turned Cygnus into a swan and gave him a permanent place in the heavens. The constellation of Cygnus is found directly overhead when crossing the meridian, and its most recognizable pattern of stars forms a well-defined cross. Its brightest star, Deneb, both marks the top of the cross (or the tail of the swan) and is also the northernmost point of the "summer triangle" asterism. Almost all of the constellation lies within the thick summer Milky Way.

Near the northeast corner of the constellation is the star cluster

M39. This is a very large, loose cluster of about 30 stars of magnitudes 7 to 10 spanning 32', or about the size of the moon at perigee. Its true appearance is best appreciated in binoculars, since in the telescope, at too high of a magnification, it becomes merely a random scattering of separate stars. It can be found 9.2 degrees ENE of the 1.2 magnitude star Deneb (Alpha (α) Cygni) or RA 21h 32.3m Dec +48° 26'. In M29, we find a much more compact cluster, and one that is almost lost in the surrounding star fields of the Milky Way. Look 1.8 degrees SSE of Gamma (γ) Cygni for a tight group (7') of 10 to 15 stars of 8th and 9th magnitude forming a shape like two parentheses turned backwards: ")(."

Another compact cluster measuring about 5' is NGC6819. Sometimes nicknamed the Fox Head, this is a rich group of stars of magnitudes 11 to 15 sitting 7.8' west of γ Cygni (RA 19h 41.3m Dec +40° 11'). It assumes the shape of a square vase or, as the name might suggest, an animal's snout.

Deneb (Alpha (α) Cygni) is worth special mention since it is one of the intrinsically brightest stars known, giving off between 60,000 and 200,000 times the light of the sun depending on its true distance. It shines at magnitude 1.2 from a distance of perhaps 3,000 light years. If placed at the standard distance of 10 parsecs (32.6 light years), it would shine at magnitude -8.7 (its absolute magnitude), while the sun at Deneb's distance would be a 13.3 magnitude weakling.

Nearby is NGC7000 (3.2 degrees E of Deneb), also known as the North American nebula. In a very dark sky, it is visible in 50mm or larger binoculars. It covers a 120' x 100' patch of the sky as a diffuse glowing cloud roughly outlining the shape of the North American continent.

Another well-known nebula is the Veil Nebula. It is the result of a supernova explosion and is separated into two parts as NGC6992 to the east and NGC6960 to the west. In 8-inch and larger scopes, this complex mass of nebulosity (which is almost 180' in diameter) is fairly evident as wispy tangles of gasses 3 degrees SSE of Epsilon (ε) Cygni. The brightest portion, NGC6992, is a 1 degree long arc south of a line connecting ε and Zeta (ζ) Cygni, while NGC6960 is dimmer, but perhaps a little easier to find, since it passes by the 4.2 magnitude star 52 Cygni. It is very much worth the effort to pan along the Veil's entire face to study the details of its structure.

Increase your power to go after a small planetary nebula about 80% of the distance from Deneb to Iota (ι) Cygni (RA 19h 45m Dec +50° 31'). Measuring a mere 2' in diameter, this 10th magnitude nebula is NGC6826. With sufficient magnification, it shows an obvious disc compared to neighboring stars.

Chi (χ) Cygni is a variable star of special interest due to its wide range of brightness. Watch for it to fade to 13th magnitude, then brighten by a factor of 3,000 to near 5th magnitude over a period of 407 days.

Even though most observers with more than just a few glimpses through a telescope are closely familiar with Beta (β) Cygni, it still deserves special mention. This is Albireo, one of the best double stars in the sky. Any telescope can split the pair because the two stars are spaced 34" apart and will show a pale golden yellow primary of magnitude 3.1 and a magnitude 5.1 companion of a wonderfully rich blue color. Albireo is easy to find as the beak or head of the swan roughly halfway between the other two members of the summer triangle, Lyra's Vega and Aquila's Altair.

Delphinus (del figh' nuss)
Abbreviation: Del
The Dolphin
Best observed between July 1st and September 30th

One of the more interesting stories relates to this small constellation on the Milky Way's edge. A Greek poet by the name of Arion, who was also a musician, traveled to Italy from Corinth to participate in a contest. He played a kithara (an instrument similar to a harp), and performed so well that he was awarded first place riches to take home with him. He chartered a ship for the voyage, but the crew attempted to throw him overboard to steal his winnings. Before they did so, however, they allowed him to play one last tune on his kithara. The beautiful music attracted a pod of dolphins in the water below, and Arion escaped the thieving sailors by jumping overboard onto the back of one of the dolphins. Thinking that Arion had drowned, the crew sailed leisurely back to Corinth, only to find that Arion, riding on the dolphin, had arrived earlier and told of the crime to the king, who sentenced the sailors to death. Another legend describes Poseidon riding on the back of a dolphin while courting the mermaid Amphitrite.

If you follow the swath of the Milky Way north from Sagittarius, you will reach Altair, the southernmost star of the familiar asterism called the Summer Triangle. From here, look northeast to the area south of Cygnus and you will see the diamond-shaped pattern of 3rd and 4th magnitude stars of Delphinus, about 3 degrees across, with its tail curving farther south.

Delphinus and Canes Venatici are the only two constellations that share a common trait of having stars named after individuals in modern times. In the case of Delphinus, the Alpha (α) and Beta (β) stars are named according to the Palermo Observatory's catalog of 1817, which designates them Sualocin and Rotanev, respectively. Spelled backwards, these two words become Nicolaus and Venator, the Latin version of the Italian name of Niccolo Cacciatore, who was the assistant to the director of the Palermo Observatory.

Touring Delphinus is best done with a high power eyepiece, since all of its objects are small and, to one degree or another, challenging. By far the easiest object in Delphinus to observe is Gamma (γ) Delphini, a double star at the dolphin's snout. Its components are of spectral types K2 IV and F8 V with a separation of 10". These two distant suns, one shining at magnitude 4.3 and the other at 5.1, appear generally white, but perhaps with a yellowish tint. At a distance of 100 light years, the brighter star shines with the equivalence of 16 suns, while the dimmer one shines with the light of 8 suns.

Turning your attention to the southern part of the constellation and peering through a fairly rich star field, you will find NGC6934. This is a globular star cluster glowing at magnitude 9 and measuring a small 4'. To find it, dial in coordinates RA 20h 34.2m Dec +7° 24.0', or look 3.8 degrees south and 5 minutes in RA east of Epsilon (ε) Delphini. With sufficient aperture, you should be able to resolve individual stars if your skies are transparent and dark.

The same cannot be said for NGC7006. At a distance of 190,000 light years, this is a very remote globular that could be an extra galactic cluster not bound to the Milky Way, but instead roaming the space between members of the Local Group of galaxies. Through even an 11-inch telescope, this cluster is difficult to discern as a globular (since its 16th magnitude stars are not resolvable), so it appears as a hazy circle 2' in diameter and glowing at magnitude 11. Instead of a globular cluster, you may think you are looking at a very distant comet that has yet to sprout a tail. The coordinates of NGC7006 are RA 21h 1.5m Dec +16° 11.0', or make an easy hop of exactly 15 minutes east in RA from γ Delphini.

NGC6905 is the most challenging target on our list to see in Delphinus. You'll want to arm yourself with perhaps a 10-inch telescope for this, a dim magnitude 12 planetary nebula sitting in Delphinus' far northwest corner (RA 20h 22.4m Dec 20° 7.0'). If you star hop to it, I suggest using 3.5 magnitude γ Sagittae as your starting point. Move 1/2 degree north and then 6.5 minutes in RA west to fifth magnitude Eta (η) Sagittae. Next, move 17 minutes in RA west to NGC6905. If you are familiar with the appearance of the planetary nebula embedded in the star cluster M46, then you will know approximately what to look for in NGC6905. But this is an even smaller version that looks like a dim star at low power, but at higher power it will become a fuzzy disk wedged between two stars that are in the neighborhood of 10th or 11th magnitude.

Draco (dray' koh)
Abbreviation: Dra
The Dragon
Best observed between May 15th and July 15th

This group of stars has represented a dragon to almost every ancient civilization that looked toward the heavens. The earliest is probably the Sumerians who saw Draco as the dragon called Tiamat, formerly a Babylonian goddess who turned herself into a dragon to give herself a fighting advantage when other gods began challenging her. The Greeks also employed the dragon in their stories of the conflicts between the Titans and the gods of Olympus. A dragon also appears in the tales involving Cadmus (the founder of the city of Thebes), the dragon that guarded the Golden Fleece, and the dragon that protected the fruits of one of the twelve labors of Heracles.

To find Draco, look toward the celestial pole. Its back forms an arch over Ursa Minor, and the head of the dragon looks toward Vega, the bright star in the constellation Lyra. If you look near the middle of the dragon's tail, you'll see Thuban, or Alpha (α) Draconis. This magnitude 3.7 star marked the celestial pole almost 5,000 years ago, but owing to precession, it has since relinquished that role to Polaris.

Draco has some excellent double stars to choose from. One of them, Nu (ν) Draconis, can be split through steadily held binoculars. Otherwise, use low power with a telescopic view to see these two white stars, both of magnitude 5 and spectral type A5, separated by a large gap of 62". Psi (ψ) Draconis is another easy double star to split, having a separation of 30". These two stars also have a similar spectral type with respect to the other (F5 and F8), so they also appear identically white, but one, at magnitude 5, outshines the other by one magnitude. Probably the most visually pleasing double star is Omicron (o) Draconis. This is also an easy pair to split (34") and to see (magnitudes 4.5 and 7.5), but what makes this duo stand out is its gold and pale blue color contrast.

The remaining targets in Draco are more elusive. Starting with the planetary nebula designated NGC6543 and nicknamed the Cat's Eye nebula, its small size (not its dimness), is what makes this object hard to distinguish from the surrounding stars. It measures no more than 20" in diameter at its outer visible edges, and its brightest central regions probably occupy an area of only about 15". It glows at magnitude 8.6, so its high surface brightness makes it stand out at high magnifications. Even so, I had to spend a few seconds adjusting the focus at 240x magnification to make sure that it didn't sharpen beyond what I saw as a lumpy, egg-shaped fuzziness. Its coordinates are RA 17h 58.5m Dec +66° 38'.

Several galaxies inhabit Draco. Unfortunately, most are too dim to see much detail after spending quite a bit of effort to find them. There are three 11th magnitude galaxies, however, that are comparatively easy to locate and are worth visiting. Starting with NGC5866 (RA 15h 6.5m Dec +55° 46' and sometimes known as M102), this shows a broad oval with flattened extensions of its disk on each side. About 1.5 degrees to the northeast is the fantastic edge-on galaxy NGC5907 (RA 15h 15.8m Dec +56° 19.5'), shining at magnitude 11.0 and, with measurements of 11.5' x 1.7', looking like the blade of a sword floating among the stars.

Finally, there is galaxy NGC6503. If the other galaxies were difficult to spot, this one should be easier, since its magnitude 10.5 light is concentrated into a smaller oval. You can find it near the base of the dragon's neck at RA 17h 49.5m Dec +70° 08.7m, or a third of the distance between Chi (χ) and Zeta (ζ) Draconis.

Equuleus (e kwoo' lee uss)
Abbreviation: Equ
The Little Horse
Best observed between August 1st and September 30th

This small, little known constellation lies between Delphinus and the nose of Pegasus and reaches its highest point in the sky at 11pm during mid to late August. Pegasus seems to be looking at his small counterpart sitting just beyond his nose. Equuleus is considered to be the invention of the Greek astronomer Hipparchus. Little mention of it is found in astronomical works since then, probably because it has only a few dim stars, and virtually no interesting objects. Myth has it that Equuleus is the horse Celeris, which is the brother of Pegasus. Depending on which story you follow, Celeris was a gift to Castor (one of the twins of Gemini) from the god Hermes, or a gift to Pollux (the other twin) from Hera, the wife of Zeus. Another story contends that Celeris sprung from the earth when Poseidon's trident struck the ground in a contest with Athena.

The constellation of Equuleus is a four-sided figure made up of stars no brighter than 4th magnitude. If you look toward Gamma (γ) Equulei, you'll see a visual binary star separated by 5.5 arc minutes. The brighter star shines at magnitude 4.7, while the dimmer one is magnitude 6. The two are of spectral types A (for the brighter star) and F, so you'll see stars of white or yellowish-white color.

For large telescopes, a dim galaxy, NGC7015, sits 1.7 degrees northwest of γ Equuleus. There are virtually no stars to use for star hopping, so luckily it's not too far away from γ. Its position is 21h 5m 37s Dec +11° 24' 50". If you have enough aperture, you'll see a dim,

13th magnitude spiral galaxy measuring 1.8' x 1.6' with diffuse boundaries getting gradually brighter toward the center.

Although not observable through an average telescope, an interesting double star is Delta (δ) Equulei, which marks the northeast corner of the box shape of Equuleus. The semi-major axis of this star is a tiny .26" with maximum separation of about .35". What sets this star apart is that its period of revolution is a comparatively short 5.7 years.

Eridanus (air id' an uss)
Abbreviation: Eri
The River
Best observed between November 15th and January 15th

This is the great celestial river that winds its way from the western foot of Orion (the star Rigel) to the southern star Achernar. (Translating "Achernar" from Arabic gives us "the end of the river.") This meandering line of stars first heads west, takes a turn to the south, flows back to the east, and finally makes its final journey southwest below the horizon. However, to be fair, from far southern US latitudes, Achernar (the 9th brightest star seen from Earth) can theoretically be glimpsed for a couple of hours when it is on the meridian, but since it only climbs a couple of degrees at most in the sky, any hill or shrub will successfully hide it from view. It can never be seen north of a latitude line connecting such cities as San Diego, Dallas, and Charleston, SC. This constellation has been seen as a river since ancient times, most often the Nile or Euphrates.

An interesting target to get us started is the triple star system Omicron 2 (ο2) Eridani. The A component (mag 4.5) is separated from the B/C pair by an expansive 83", while B and C (magnitudes. 9.7 and 10.8, respectively) stand about 9" apart. The group is only 16 light years from Earth, making the brightest star the 8th nearest of the naked eye stars. B is an important object simply because of what it is: a white dwarf. It's one of the most easily observable stars of its kind. The 0.44 solar masses it contains is squeezed by gravity into a body that is less than 2.5 times the diameter of Earth, making its density 90,000 times that of water. A golf ball sized chunk of this star would weigh close to 3.5 tons. The C star is a red dwarf and a featherweight among its stellar counterparts, weighing in at only 0.2 solar masses.

Another double star important for its visual appearance is Theta (θ) Eridani. Easily seen with the naked eye at magnitude 2.9, it appears through a telescope as a pair of brilliant white stars of magnitudes 3.5 and 4.5, separated by about 8.5".

There are dozens of galaxies in Eridanus, but three that take center stage are NGC1300, NGC1232, and NGC1291. The first two happen to be close to each other, forming a triangle with Tau4 (τ4) Eridani. NGC1300 is a barred spiral of magnitude 11.2 and measuring 6.2' x 4.1'. You may not notice the two spiral arms, but you should see the nucleus and bar making a thin oval structure. NGC1300 is found 2.4 degrees north of τ4, or RA 3h 19.7m Dec -19° 24.7".

Moving 2.6 degrees to the southwest will reveal NGC1232, a slightly brighter galaxy of magnitude 10.6 and larger than NGC1300, covering 7.4' x 6.4'. This one has closely wrapped multiple arms that seem to merge with each other to give this galaxy the appearance of a soft circle of light. You will find NGC1232 at RA 3h 9.8m Dec -20° 34.9'.

Near the far southern regions of the constellation is NGC1291, another bright spiral galaxy with tightly wound arms. It glows comparatively brightly at magnitude 9.4, and it is significantly larger than the other two galaxies, measuring 9.7' x 8.1'. However, it sits well down in the sky toward the southern horizon 3.7 degrees ESE of our double star θ Eridani. Any significant dust or haze in the air will compromise your view of this galaxy. Look at RA 3h 17.3m Dec -41° 6.5' for this object.

A fairly easy deep sky object to view is NGC1535. It's a planetary nebula of magnitude 10 and 20" in size. Move 5.5 degrees south of the triple star o2, or zoom in to RA 4h 14.2m Dec -12° 44.5'.

Fornax (for' nacks)
Abbreviation: For
The Furnace
Best observed between November 1st and January 30th

Many of the faint constellations were created by the German astronomer Johannes Hevelius, but the credit for this particular group of stars goes to the Frenchman Nicolas-Louis de Lacaille, who studied astronomy in the 18th century. Originally he named it Fornax Chemica, or the chemical furnace. The neighboring constellation of Eridanus winds its way around Fornax, which you can find between 2 and 4 hours of RA and -40 and -25 degrees declination. Maybe the best way to locate Fornax is to orient your view with respect to the constellations Orion and Cetus. Make a line from Alnitak, or the easternmost star of Orion's belt, follow it through Rigel (Orion's western foot), and continue to a point directly below the head of Cetus. The scattered stars of Fornax trace out no particular shape and shine dimly between magnitudes 4 and 5.

Near Fornax's border with Eridanus is this constellation's main draw. Here is a cluster of 23 galaxies, ten of which are visible in the same eyepiece if your scope is capable of a very wide field of view (1.5 degrees). The ones that stand out in this compact group are as follows:

NGC1399: Magnitude: 9.8. Size: 6.9' x 6.4'. Position: RA 3h 38m 28s Dec -35° 26' 58". Here is a bright elliptical galaxy with a bright core and diminishing haze surrounding it.

NGC1365: Magnitude: 10.1. Size: 11.2' x 6.2'. Position: RA 3h 33m 36s Dec -36° 08' 17". A remarkable double-armed barred spiral.

NGC1380: Magnitude: 11.0. Size: 4.7' x 2.3'. Position: RA 3h 36m 27s Dec -34° 58' 33". A lenticular galaxy with a bright center.

Outside of this group are two other bright galaxies of note. Two and one-third degrees southwest of NGC1365 is NGC1316 (magnitude 9.2, 11.9' x 8.5', RA 3h 22m 42s Dec -37° 12' 28"). This is another lenticular galaxy with a bright core surrounded by a haze of innumerable stars.

Farther north in the center of the constellation is NGC1097. Shining comparatively brightly at magnitude 9.9, this galaxy is found 2.2 degrees north-northwest of Beta (β) Fornacis (or RA 2h 46m 19s Dec -30° 16' 21"). If the sky is dark and your aperture generous, look for the bar spanning the galaxy's center.

Tucked away in the northeast corner of Fornax is NGC1360, a planetary nebula that offers us a welcomed change of pace from the wealth of galaxies. This oval-shaped nebula measures 6.5' in its long axis with an easy to spot 8th magnitude central star. Find it at RA 3h 33m 18s Dec -25° 51' 00". Another way to find it is to point your telescope northeast of Alpha (α) Fornacis, where you will see it forming a straight line about equally distant from this star as Beta (β) Fornacis is to the southwest.

Gemini (jem' in eye)
Abbreviation: Gem
The Twins
Best observed between January 1st and March 30th

To the Chinese, they were yin and yang, the dual forces of nature. In ancient Rome, the pair was sometimes seen as Romulus and Remus, who were the founders of that city. Classical mythology gives us the traditional names of the two bright stars that represent the twins Castor and Pollux, who were hatched from an egg created by the seduction of

Leda by Zeus while he was in the form of a swan. After Castor and Pollux were raised by Chiron, a centaur, they joined Jason and the crew of the Argo on his journey to find the Golden Fleece. The twins helped to calm a terrible storm that they encountered while at sea, and as a result, became known as protectors of sailing vessels.

Gemini stands above Orion's left shoulder with the twin's feet within the winter Milky Way. If you ever confuse the two stars of Castor and Pollux (Alpha (α) and Beta (β) Geminorum, respectively), use letter association and just remember that Castor is nearer to Capella (in Auriga), and Pollux is nearer to Procyon (in Canis Minor). Or perhaps consider that Pollux has a pinkish tint (since it is a K type star) while Castor is of a bluish white type A. At first glance, Castor may appear as a single star, but it is actually a system of six stars divided into three pairs that can be split with just about any telescope. The closest pair are 2" apart, and so require magnification of perhaps 180X or more, and appear like two brilliantly white beacons of magnitudes 2 and 3. The third pair appears as a 9th magnitude star 73" from the main quartet. None of these pairs can be resolved individually, since they sit so close to each other.

Due to Gemini's proximity to the Milky Way, the constellation is rich in star clusters and nebulae, but lacks any significant galaxies. The most impressive star cluster is the 5th magnitude open cluster M35. Located at RA 6h 08.8m Dec +24° 20', or approximately 2.5 degrees to the northwest of Eta (η) Geminorum, this cluster, about 2,800 light years distant, contains about 120 stars in an area spanning half of a degree. Just on the edge of M35 is a small open cluster, NGC2158, that is easy to overlook if you become lost in the splendor of M35, but you will see a separate fuzzy patch of less than 10' in size that just overlaps M35 on its southwestern limb. Two other similarly sized open clusters are nearby. IC2157 is about 2' west of NGC2158, and NGC2129 is about 3' west of 1 Geminorum. These small clusters are a nice challenge, even under dark skies.

An interesting planetary nebula resides in Gemini one degree south and 9' east of Delta (δ) Geminorum. Designated NGC2392, the popular name of this object is the Eskimo Nebula because of the way it resembles a face wrapped in a fur-lined hood in photographs. Through a telescope, however, don't expect to see much more than a faint 8th magnitude fuzziness about 45" wide at RA 7h 29.2m Dec +20° 55'.

If you find yourself observing Gemini in December, remember that the Geminid meteor shower peaks on the 13th, with the radiant very near Castor. These 22 mile-per-second meteors may produce 15 to 20 bright meteors and up to 40 additional fainter ones per hour.

Grus (groos)
Abbreviation: Gru
The Crane
Best observed between September 15th and October 15th

This constellation is comparatively recent, appearing in Johann Bayer's atlas in 1603. Originally, the stars of Grus were part of Pisces Austrinus (the Southern Fish), but since then they have been grouped into a separate, distinct constellation. There is little history associated with Grus, since it contains stars of 2nd magnitude and fainter, and it lies in a far southern declination. Therefore, it largely escaped the attention of the ancients. To see Grus in its entirety, you'll need to live south of 33 degrees north latitude, since Grus' southern boundary runs along declination -56.5 degrees. To find it, look for the 1st magnitude star Fomalhaut (the brightest star in the constellation Pisces Austrinus) sitting alone in the southern sky. Grus is straight south of Fomalhaut assuming the shape of an upside down, broken cross, or the crane it is supposed to be with its neck tilting to the west.

Grus is clear of the obscuring Milky Way, allowing us to look past our home galaxy into intergalactic space. As a result, Grus is home to a wealth of galaxies and none of the comparatively nearby nebulae and star clusters. On our way to some of these galaxies, let's first stop at a visual pair of stars near the heart of Grus. The two, Pi 1 (π 1) (magnitude 6.4) and Pi 2 Gruis (magnitude 5.6), sit within the triangle of stars formed by Delta (δ), Alpha (α), and Beta (β) Gruis. They are separated by 4.3 arc minutes, so they can be split with just binoculars. Each star, however, presents more than initially meets the eye. Zooming in to π 2 reveals that this star is a double itself. It has a dim 10.6 magnitude companion 11" away at a PA of 212.5 degrees. Moving to π1 to the west, we find that this star is an irregular variable fluctuating between magnitudes 5.4 and 6.7. Comparing π 1 to π 2 gives us a good way to judge the variable's behavior.

If you'd like to find another variable star, try T Gruis. This one is more predictable, being a long period variable with a fairly short period (for the type) of 136 days. At maximum, it is a not too bright 7.8 magnitude. At minimum, it all but disappears to magnitude 12.3. You can find T Gruis at RA 22h 25m 41s Dec -37° 34' 08", 1.7 degrees NNW of Nu (ν) Gruis.

The galaxies in Grus are quite dim. You won't find an M31 or M81 caliber galaxy here. However, some are well within the grasp of medium sized telescopes. Up in the northeast part of the constellation is NGC7410, a spiral tilted almost edge on. It glows with a magnitude of 11.5 and measures 5.3' x 1.6'. The center of this galaxy is quite bright and broad, with the spiral arms fading rapidly toward the edge. NGC7410 sits at RA 22h 55m 00.7s Dec -39° 39' 41" in a line with Delta (δ) and Rho (ρ) Gruis, 2.8 degrees northeast of ρ.

One and a half degrees south southeast of 7410 is NGC7424. This face-on spiral galaxy glows more brightly at magnitude 10.9, but the problem with spotting this one is that this light is spread over a larger area of 9.5' x 8.0'. Thus the surface brightness is significantly lower. Its core is the most noticeable feature, with its loose arms reaching faintly beyond. Find NGC7424 at RA 22h 57m 18.0s Dec -41° 04' 09".

NGC7496 is easy to find. Just move half a degree east of Theta (θ) Gruis and you'll see a compact 11.9 magnitude fuzzy spot measuring 3.3' x 3.0'. For those of you who do want the specific location, it is RA 23h 09m 47s Dec -43° 25' 43".

Swinging completely over to the southwest corner of the constellation we will run across NGC7144, a galaxy that shows a different character. Evidently, here is an elliptical galaxy with a diameter of 3.7'. There is no detail to this one except for a perfectly round, evenly lit glow from countless stars. The trouble we have here is that it lies so low in the sky that we may have to compete with atmospheric interference and the limited time it is above the horizon. NGC7144 is 2.9 degrees southwest of α Gruis, or RA 21h 52m 43s Dec -48° 15' 16".

Hercules (her' kyoo leez)
Abbreviation: Her
The Strongman
Best observed between May 15th and July 30th

From Greek myth we get the elaborate story of Hercules, who was the son of Zeus and a mortal princess. Hera (the wife of Zeus), in her jealousy, sent serpents to kill the infant Hercules, but his strength was immediately displayed when he strangled the snakes. When that didn't work, Hera managed to send him away to become a servant to King Eurystheus. There he grew up to be stronger than any mortal man. In order to be released from the servitude of the king, he was commanded to complete the Twelve Labors, including slaying many terrifying creatures such as the Nemean Lion (which became Leo) and a several headed monster called Hydra (also placed in the sky).

The demise of Hercules came after he had won the affection of Deianeira, a beautiful maiden. When Nessus (a fearsome centaur), kidnapped Deianeira, Hercules mortally wounded him with an arrow. Nessus gave Deianeira a drop of his blood and deceived her by saying that Hercules would be forever in love with her if he ever came in contact with it. But when Hercules wore a tunic that Deianeira had dotted with the blood, his flesh began to burn away. As a result of this blunder, Deianeira's emotional anguish and Hercules' physical pain led them to both kill themselves. Zeus put Hercules in the sky for all the world to see, and to remember the mighty hero.

Hercules is still low in the sky after dark during the middle parts of May, and reaches the meridian just after midnight. By the end of July, he is past the meridian and moving toward the western horizon just as the sky gets dark. He poses in a kneeling position to the west of the bright star Vega in the constellation of Lyra. The most recognizable part of Hercules is the quadrilateral of stars that form the "keystone" figure. The other stars of Hercules forming his limbs splay out from this keystone and are in the range of 2nd, 3rd, and 4th magnitude.

Hercules is home to one of the best deep sky objects visible from this hemisphere, the great Hercules globular cluster M13. When Charles Messier discovered it in 1764, he described it as "a nebula which I am sure contains no star." This gives you an idea about the kind of instruments he had at his disposal in his day. It will rise to within a few degrees from the zenith and will appear as a diffuse spot of magnitude 5.9 to the naked eye on nights of exceptional darkness and clarity. A small telescope of about 4 inches will show this spot as a large ball with a few of the brighter stars coming into view. Eight inches or more of aperture will make this globular cluster bloom into a dense explosion of hundreds of stars while the other hundreds of thousands fill in the spaces in between with a dense glow spreading across 16' of the sky. A prolonged gaze at M13 will be a treat for your imagination as geometrical patterns of stars will come in and out of view. Besides the brightness of M13, the other aspect that makes this object a joy to view is how easy it is to find. Simply look 2.5 degrees south of Eta (η) Herculis, the 3.5 magnitude star marking the northwest corner of the Keystone asterism. If you wish to use your setting circles or a go-to scope, M13's location is RA 16h 42m 53s Dec +38° 55' 20".

The dazzling nature of M13 makes us forget that a galaxy lies less than half a degree from it. However, a large telescope is required due to its weak 12.2 magnitude glow. NGC6207 is very much overshadowed by M13, but presents a challenge for those who want to hunt it down. Look 27 arc minutes to the northeast of M13 for this small, oblong glow, or RA 16h 47m 59s Dec +36° 49' 59".

While M13 is an unquestionable visual treat, don't forget about Hercules' other great globular cluster, M92. This spectacle is only slightly smaller and dimmer than M13. If you draw a line from η to Iota (ι) Herculis, M92 is about 2/3 of the way between the two, or RA 17h 17m 6s Dec +43° 8' 00". Robert Burnham, Jr., in his famous Celestial Handbook, indicates that M92 holds its own among the sky's other globular clusters, indicating that few clusters of its type compare to its beauty. M92 shines at magnitude 6.5 and spans an area of 11.2 arc minutes.

From these ostentatious objects we can also seek some of the more obscure things Hercules has to offer. Toward the south, a little less than 8 degrees from the "keystone," is the small planetary nebula NGC6210 (RA 16h 44m 30s Dec +23° 49' 00"). Boost the power for this tiny nebula since its diameter spans only about 12 arc seconds. But the small size, coupled with its fairly bright 9th magnitude glow, makes it fairly easy to spot.

Alpha (α) Herculis is named Ras Algethi from the Arabic phrase that means "head of the kneeler," referring to Hercules' kneeling pose. It is a naked eye irregular variable star. Its brightness can be compared to nearby Delta (δ) Herculis, shining at magnitude 3.1. α goes from that same magnitude down to about magnitude 3.9 over a period averaging about 90 days. It has a spectrum of M5, which gives it a reddish color.

Hydra (high' druh)
Abbreviation: Hya
The Sea Serpent, or Female Water Snake
Best observed between March 1st and April 30th

Hydra plays several roles in mythology. One portrays Hydra in a battle with Hercules, who was able to defeat the multiple-headed monster by cutting off each head and then cauterizing the wounds with a torch to prevent another head from immediately growing back. A more meek Hydra appeared in a story involving Corvus, the crow. Corvus had a task to bring back some spring water for Apollo, but spent some time next to a fig tree waiting for the fruit to ripen. After eating some figs, Corvus finally returned with some water along with Hydra, using the story that Hydra had attacked him and thus was the reason for his delay. Myth also says that Hydra was a fearsome giant who fought with the gods of Olympus and was finally killed by one of Zeus' thunderbolts.

To observe Hydra in its entirety is to spend most of the night under its stars. Its vast size makes it stretch almost 100 degrees from end to end, or more than a quarter of the celestial sphere. Its head is south of Cancer (the crab) and its body curves its way through the heavens past Crater, Corvus, and Virgo to finally end with its tail bordering Libra. Its name suggests a feminine gender, and is not to be confused with the southern constellation Hydrus, or the male water snake.

A logical method of observing Hydra is to go from west to east, and the first object to view is also one of the most satisfying. It's M48.

Tucked against the western boundary of the constellation is this galactic cluster of large proportions. It glows with a magnitude of 5.8 and covers about 50'. Through binoculars, it appears as a cloudy area with a few embedded bright stars. In the telescope, you'll see an expansive and loose collection of stars filling the full field of view of your eyepiece at low power. It is located in a rather blank area at RA 8h 13.8m Dec -5° 48.0'. If finding it without setting circles, it will easily be visible through your finder scope about 11 degrees southwest of the serpent's head.

Looking east we will find NGC3242, a planetary nebula with an intriguing nickname: the "Ghost of Jupiter." It is fairly bright at magnitude 9 and a diameter of 16", so most telescopes can capture it well. Many observers are struck by its blue color. An 8-inch scope is all that's necessary to show an inner, elliptical structure surrounded by a circular halo, all set aglow by the magnitude 12 central star. Large instruments in the 12-inch neighborhood may show a layered structure within the brightest parts of the nebula. Jupiter's Ghost can be found near the center of the constellation 1.8 degrees south of magnitude 3.8 Mu (μ) Hydrae (or RA 10h 24m 49s Dec -18° 38' 15".)

Patience rewards those who watch the full length of Hydra rise. The meridian bisects the constellation by 11pm during the middle of March and we see more deep sky wonders in Hydra's eastern half. M68 is a globular cluster located south of Corvus. You can find it in a couple of ways. Follow the line formed by Delta (δ) and Beta (β) Corvi going south. About half the distance that is between these two stars past β is a 5.4 magnitude star. M68 is half a degree to the northeast from this star. Otherwise, look about 2/3 the distance from Zeta (ζ) to Gamma (γ) Hydrae or zoom in at RA 12h 39.5m Dec -26° 45.0'. M68 is an easy globular at magnitude 8.2 forming a ball 12' in diameter. Its 13th magnitude stars are resolvable in 8-inch and larger telescopes, and smaller instruments will see a mottled glow.

The role of Hydra's showpiece object can be given to M83, a bright spiral galaxy forming a triangle with γ and Pi (π) Hydrae. M83 is to the southwest of a line drawn between these two stars and about equally distant from them (or RA 13h 37m Dec -29° 52.1'). This is one of the brightest galaxies in the sky. Its magnitude 8.1 light is spread across a size of 12.9' x 11.5'. The galaxy's two bright arms are first noticed, but a closer look reveals a third, dimmer arm that have caused some to call this galaxy a 3-armed spiral. It presents its full face to us, and many clumps of nebulosity and star clusters have been observed throughout the arms. There is an unusual amount of hot, giant stars in this system, and many supernovae have occurred here with a frequency about 15 times the average.

Looking above M83 is the long period variable star, R Hydrae. It is located 2.5 degrees east of γ Hydrae. Along with Mira, this variable was one of the earliest to be noticed. This one was first measured in the 17th century when its light curve spanned about a year and a half. Since then, its period has shortened to 389 days as it widely pulsates between 3.5 and 10.9. As is typical with the Mira type variables, its fluctuations may return to an 18-month interval in a few centuries time.

Lastly, look to the eastern edge of the constellation to NGC5694. This very distant globular cluster shines across space at magnitude 10.2, and is 3.6' in size. It's located 2 degrees southwest of a curved pattern of the four stars of 54, 55, 56, and 57 Hydrae which range between magnitudes 5 and 7. NGC5694 will appear as a round glow with some mottling present. (RA 14h 39.6m Dec -26° 32'.)

Lacerta (lah ser' tuh)
Abbreviation: Lac
The Lizard
Best observed between July 15th and October 31st

This group of stars was largely neglected by the ancients since only dim stars comprise its indiscriminate shape. It wasn't until late in the 17th century that astronomers began concocting names for this scattering of stars. One of the original ones comes from the French astronomer Augustin Royer, who proposed the ponderous name of "Scepter and the Hand of Justice," in honor of Louis XIV. Johann Bode of Germany suggested the name "Frederici Honores" after the reign of Frederick the Great of Prussia. However, it was Johannes Hevelius, also German, who is given credit for the name Lacerta, which has endured to the present day and left the other names to be forgotten over time.

Lacerta has a northerly spot in the sky, sitting between Cygnus and Andromeda beneath (or north of) the feet of Pegasus. Look east of Deneb, the brightest star of Cygnus, and you will see the zigzag of stars shining between magnitudes 3.5 and 4.5 that form the body of Lacerta. It straddles the meridian just after 1am in August, and closer to 11pm in September.

This constellation would be especially uninteresting if it didn't hold the position it does in the sky. It's just on the edge of the northern summer Milky Way, meaning that Lacerta is full of wonderful galactic star clusters.

The two largest ones of these are near the border of Cygnus. NGC7209 is a bright magnitude 6.7 cluster spread across 25' of the sky. You'll find it 16' in RA west of 4.5 magnitude 2 Lacertae. It holds about 50 stars ranging in magnitude 9 to 12. A bit farther north is NGC7243. Being about the same size (21') and brightness (mag. 6.4) makes it a near twin to NGC7209. However, not quite as many stars populate this cluster. Look closely and you'll see a double star within this group. They are a pair of hot A0 type stars separated by 9" and each shining at magnitude 8.5.

Floating near the lizard's head, like swarms of insects in danger of being devoured, are three modest clusters that are all small and dim but, nevertheless, worth your time if you haven't observed them before. NGC7296 near Lacerta's snout is a mist of stars 4m 40s in RA east of the 4.4 magnitude star Beta (β) Lacertae. This cluster measures a meager 4.0' in diameter, but is fairly rich. IC1434 is another rich cluster, 8.0' in size, and of magnitude 9.0. It shows a branched structure by its stars of 12 to 15 magnitude. The stars of the Milky Way collect into the tight knot of NGC7245, the third cluster in this area. This one sits near the northwest edge of the constellation 4.5 degrees due north of NGC7243. Below is a list of the six best deep sky objects found in Lacerta.

Object	Mag	Size	Location	
NGC7209	6.7	25.0'	RA 22h 5.2m	Dec +46° 30'
NGC7243	6.4	21.0'	RA 22h 15.3m	Dec +49° 53'
NGC7296	10	4.0'	RA 22h 28.2m	Dec +52° 17'
IC1434	9	8.0'	RA 22h 10.5m	Dec +52° 50'
NGC7245	9.2	5.0'	RA 22h 15.3m	Dec +54° 20'
IC5217	13	6.0"	RA 22h 23.9m	Dec+50° 58'

Lacerta's real challenge comes in the form of planetary nebula IC5217, which is situated on a line almost midway between β and 4 Lacertae (or RA 22h 23.9m Dec +50° 58.2'). Almost stellar in appearance, IC5217 only reveals its 6" disk at high magnification. It glows faintly at magnitude 13.

Leo (lee' oh)
Not abbreviated
The Lion
Best observed between February 15th and May 15th

Lying on the ecliptic and making its appearance just above the horizon as the sun sets in March, but well past the meridian as the sky darkens in May, is one of the more recognizable constellations of the zodiac, Leo the Lion. The vast majority of ancient civilizations indeed saw a lion in this collection of stars. The exceptions include the Chinese, who saw a horse, and the Incas, who thought the stars outlined the figure of a puma. Other than its regal shape, one reason for a lion being connected with the constellation is the observation that lions came to the Nile to cool themselves when the sun was passing through Leo during the hot days of late summer. The Greeks considered that Leo specifically represented the mythological Nemean Lion which terrified Corinth until Heracles came to the rescue.

Leo contains several bright stars, so finding it in the sky is simple. If you are familiar with the pointer stars of the Big Dipper that lead you to Polaris (these are the outer two bowl stars that are opposite the handle of the dipper), just follow that line formed by those pointers in the opposite direction from Polaris and your eyes will land firmly on Leo's back.

Gamma (γ) Leonis (named Algieba) provides a good first target to view. It is an excellent double star whose components shine at magnitudes of 3.3 and 3.5 with an identical yellow-orange color. It will take high magnification to overcome its close separation of 4.5", and little change will occur as it slowly increases toward about 5" by the year 2075. γ is the brightest star in the curve of stars forming Leo's head and shoulder, often seen as a sickle.

Sitting 9.5 degrees northeast of Algieba is 54 Leonis. This double star is a close copy of Algieba in that it has an almost identical PA and a similar difference of magnitudes between the two (4.5 and 6.5), but they have a slightly wider separation of 6.5" and a color of white or slightly blue.

One of the best variable stars to observe is R Leonis. It is easily found 21' in RA (or just over 5 degrees) west of Regulus near 5.6 magnitude 18 Leonis and 6.3 magnitude 19 Leonis. R is a long period variable of the Mira class, pulsating from minimum near 10th magnitude to a maximum of about 5.5 magnitude in a period of 156 days. It forms one point of a triangle south of 19 Leonis with two 9th magnitude stars, and it's easy to determine which one it is by its red color if it is closer to its dim phase.

Galaxies in Leo are too numerous to mention in detail. There are several, however, that deserve special attention. Starting 2.5 degrees southeast of Theta (θ) Leonis is the trio of galaxies M65, M66, and NGC3628. M65 and M66 both measure 8' x 4' and appear as parallel strips south of the more distended NGC3628, which is more or less perpendicular to the Messier pair. All are in the neighborhood of 10th magnitude and can be framed in the same field of view in a low power eyepiece centered at RA 11h 20m Dec +13° 15.5'. Half the distance between these and Alpha (α) Leonis (Regulus), is another trio of bright Messier galaxies: M95, M96, and M105. M95 is the dimmest of these three at magnitude 11.5 but, being a barred spiral, is perhaps the most photogenic. M96 is a tight spiral with a bright central bulge, while M105 is a bright elliptical of magnitude 10.1.

An isolated galaxy, NGC2903, can be found 1.5 degrees south of Lambda (λ) Leonis. Its bright 9.6 magnitude glow and fairly large size of 12.5' x 6' make it an easy target. NGC2903 is located at RA 9h 32.2m Dec +21° 30'.

Leo Minor (lee' oh migh' ner)
Abbreviation: LMi
The Little Lion
Best observed between February 15th and May 15th

If you look to the well-known and shapely constellation of Leo, the lion, and then move your gaze slightly north toward equally impressive Ursa Major, you'll come to one of the least interesting and least inspiring constellations to the eye. That's because the brightest star of Leo Minor, Beta (β) Leonis Minoris (Alpha has somehow been dropped), is only magnitude 3.8, and the others are 4.5 or less. In fact, Leo Minor sits in an area of the sky where the ancient Greeks gave the name amorphotoi, meaning "undeveloped." Johannes Hevelius came along in the late 1600's to give these stars their present name to fill in this comparatively empty part of the sky.

Leo Minor is a fair distance from the winter Milky Way, so what we expect to find here are no star clusters, and a good amount of galaxies and stellar objects. Looking to the west side of the constellation we find R Leonis Minoris, a cool M type star that is also a long period variable (LPV) pulsing from magnitude 6.3 to a virtually invisible (depending on your telescope) 13.2 over a period almost matching a year (372 days). R is roughly 4.5 degrees west of 21 Leonis Minoris or, specifically, RA 9h 45m 35s Dec +34° 30' 45".

There are several galaxies in Leo Minor. Most are small and in the vicinity of 12th to 13th magnitude. A couple are brighter and within the grasp of most backyard telescopes. We find one of these galaxies, NGC3344, at RA 10h 43m 31s Dec +24° 55' 25". It's one of Leo Minor's brightest ones at magnitude 10.7. It spans an area of 7.0' x 6.5'. NGC3344 is a wonderful face-on spiral galaxy with a sharp nucleus and diminishing spiral arms. Its general orientation with respect to the Earth is similar to the nearer galaxies M33 in Triangulum and M51, the Whirlpool Galaxy, in Canes Venatici. While not as large as these two other galaxies, NGC3344 rivals their appearance nonetheless, if we allow for the greater distance.

NGC3486 is of similar brightness (magnitude 10.7) a bit farther east at RA 11h 00m 24s Dec +28° 58' 33". This galaxy is of a very similar appearance, also being face-on to us and a size of 7.0' x

5.2'. You'll see a galaxy with a bright, broad nucleus.

Those of you with larger aperture can try for the interacting pair of galaxies NGC3396 and NGC3395 1.4 degrees southwest of 46 Leonis Minoris. Both of these are of magnitude 12.5 and a small 3.0' x 1.2' for NGC3396 and 2.1' x 1.2' for NGC3395. The centers of these galaxies are only 1.7 arc minutes apart, and long exposure images show distinct nuclei. However, their adjacent edges are being smeared together from gravitational interaction. You can find the pair at RA 10h 49m 53m Dec +32° 59' 10".

Lepus (lee' puss)
Abbreviation: Lep
The Hare
Best observed between December 1st and February 28th

The winter skies are dominated by many attention grabbing constellations which are home to some very bright stars and star clusters. But lying below Orion, the constellation that is arguably guilty of stealing the most limelight, is a rather unimposing constellation by the name of Lepus. This collection of stars represents a rabbit or a hare, and its location here is significant for two reasons. Firstly, consider that Orion has been seen as the Sun god, and the hare has been connected to the moon by the fact that many cultures have seen the shape of a hare in the patterns formed by the lunar maria. Therefore, just as the sun chases the moon across the sky, Orion acts as the hunter, and Lepus is his prey, having been snared and lying at Orion's feet. Secondly, hares also serve as prey for eagles, so being afraid of Aquila, the eagle, Lepus is seen safely rising in the east as Aquila sets out of sight in the west.

To begin your observation of Lepus, look for the globular cluster M79. Among the globulars in Messier's catalog, M79 is one of the smaller ones similar to M70 and M54 in Sagittarius and M72 in Aquarius, but judging it on its own merits makes it a wonderful sight in any telescope. Glowing at magnitude 8.3, you'll see it at RA 5h 24.2m Dec -24° 31', or 3.8 degrees south and 4 minutes west of Beta (β) Leporis. Use medium to high power to see this globular the best.

While you're in the neighborhood, slide about 2/3 of a degree to the southwest to a star designated Herschel 3752. At low power, this will appear as a double star with a wide separation of 59", but looking more closely with a high power eyepiece will turn this binary into a triple system. Its components are of magnitudes 5.5, 6.4, and 9.0 with the closer pair only 3" apart. They sit in almost a straight line with PA's of 97° and 106° with respect to the primary.

Tucked near the northwest corner of Lepus is a rare, high mass star in the late stages of evolution and falls outside of the mainstream stellar categories. Designated R Leporis, it is a type of star called a carbon star, so termed because it uses carbon as a catalyst for its nuclear reactions instead of the simpler hydrogen fusing mechanisms used by lower mass stars. Like the Garnet Star in Cepheus, R Leporis has a similar name, the "Crimson Star," due to its red color. It is actually a variable star swinging between magnitudes 5.5 and 10.5, and the richness of its color likewise varies. While at its dimmest, its redness stands out like a neon bulb, but near its brightest, it is more coppery. When looking at R Leporis, you can simply enjoy its appearance, or contemplate the unusual nuclear activity taking place inside as a last ditch effort to keep the star shining.

For those wishing challenges, Lepus offers IC418, a very small 11th magnitude planetary nebula at RA 5h 27.8m Dec -12° 31.1', or start at Nu (ν) Leporis and move south 1/3 of a degree and then east 8 minutes. At 120x through an 8-inch scope, I was just able to see this nebula as a fuzzy disk compared to the surrounding stars. With decent surface brightness, this nebula can be seen fairly easily with sufficient magnification.

NGC1964 and NGC1744 are two faint galaxies that can be elusive depending on seeing conditions and the type of telescope used. NGC1964, at magnitude 11.6, is the brighter of the two and can be found at RA 5h 33m Dec -21° 52', or 1.2 degrees south and 5 minutes east of Beta (β) Leporis. NGC1744 resides at RA 5h 00m Dec -26° 05' or 3.6 degrees south and 6 minutes west of Epsilon (ε) Leporis. Both of these galaxies don't appear much more than dim, oval smudges, but some satisfaction results from finding them.

Libra (lee' bruh)
Abbreviation: Lib
The Scales
Best observed between May 1st and June 30th

Libra is unique among the constellations of the zodiac in that it is the only one that does not represent a living thing. Originally, the stars of Libra were included in the adjacent constellation of Scorpius, but the ancient Romans and Egyptians considered them to outline a separate object on their own. They saw them as a scale, or balance, since the sun was in Libra at the time of the autumnal equinox when the length of night and day are equal. We keep this identification with a scale today, but since then, precession of the Earth's axis has taken the location of the sun at the beginning of autumn to western Virgo. By the year 2440, the sun will be visiting Leo at this same time of the year.

You will see Libra in the space between Scorpius and Virgo where the ecliptic dips south. The constellation is on the meridian at 1am at the start of May, and 9pm around June 30th.

Libra's two brightest stars have the wonderful names of Zubenelgenubi, or "southern claw," and Zubeneschamali, or "northern claw." The translation from Arabic indicates this pair's former association with neighboring Scorpius. Zubenelgenubi, Alpha (α) Librae, is an optical double star separated by a very wide 231", so this makes it a suitable target for binoculars. The primary shines at magnitude 2.7, and its companion, 5.2. The spectral types are A3 and F5, respectively. The two stars show the same proper motion through space and imply that they are a true physical pair. If so, their optical separation as seen from 75 light years away (as measured for the primary) makes for a true separation of about 4,800 AU (446.4 billion miles), or a speed of light trip of almost 4 weeks from one star to the other.

An Algol-type eclipsing binary star can be found in the form of Delta (δ) Librae. It has a period of just under 2 days 8 hours and can be seen dimming from maximum to minimum in about 5.5 hours as the dimmer component, a star that is about 3 times brighter than the sun, passes in front of an even brighter star, which shines with the light of 46 suns. There is a magnitude 6.5 star 19' to the northeast of δ that you can use for comparison. δ normally shines at magnitude 4.8, but dims to 5.9 while in eclipse.

The one deep sky object in Libra that most small to moderately sized telescopes can capture is the globular cluster NGC5897. Don't expect to see anything like an M13 or M15, however. This is a sparsely packed collection of stars spread out over an area of 12.6' and shining at magnitude 8.6. Its appearance is diminished by the 40,000 light years that separate it from the observer, and by the fact that it is not too far from the thickest parts of the summer Milky Way, so we are required to peer through the outskirts of our galaxy's obscuring dust clouds. At first, it may take you a moment to realize that you have actually found it. Its soft glow is barely apparent against the dark background, and at high magnification may look like a dense galactic cluster rather than a globular. NGC5897 can be found at RA 15h 17.4 m Dec -21° 01', or 1.7 degrees southeast of Iota (ι) Librae.

It's unfortunate that peering even deeper into space in this area doesn't get any easier. There are close to twenty galaxies in Libra on a detailed star chart, but they are all very dim and elusive. One of the more comparatively bright galaxies can be found very near Zubeneschamali, or Beta (β) Librae. This is NGC5885, a faint spiral of magnitude 12 sitting 51' southwest of this star. Its coordinates are RA 15h 15.1m Dec

-10° 5.2'. A dark sky and generous aperture are necessary to spy this distant galaxy. It shouldn't present anything more than a smudge with dimensions of 3.6' x 3.1', but there is some satisfaction in having found a galaxy in Libra.

Lupus (loo' puss)
Abbreviation: Lup
The Wolf
Best observed between May 1st and June 15th

The entirety of this southern constellation just clears the horizon from latitudes south of 33 degrees north. It appears as a sprinkling of medium-bright stars southwest of the curving tail of Scorpius. Its modern image is that of a wolf, but in ancient times it was called Therion, an unspecified wild animal.

The first object in Lupus that comes into view is NGC5824, a globular cluster in the northwest corner of the constellation. This globular is both easy and difficult to find. Or a better way of putting it is to say that it is easy to see, but difficult to recognize. It is fairly bright at magnitude 9.0 with a size of 6.2'. I was scanning its area with an 11-inch telescope and wondering why I couldn't see it until I realized that it was masquerading as a star. The center of this globular cluster is so densely cluttered with stars that, at low power, it looks stellar. At high power, the center becomes extended with a faint haze surrounding it. Unfortunately, no stars are resolvable except perhaps through the largest of backyard telescopes, but its bright nucleus is quite impressive. About 10.5 degrees north of Beta (β) Lupi (or the width of your fist held at arm's length) is a 5.5 magnitude star, and 1/2 degree SSE of this is NGC5824. More precisely, this globular is at RA 15h 04m Dec -33° 04'.

Moving 10 degrees to the southeast in the wolf's neck we find another globular that couldn't be more different than NGC5824, as there is no mistaking what this object is. This one is NGC5986, 2.8 degrees WNW of Eta (η) Lupi, or RA 15h 46.1m Dec -37° 47'. Measuring a large 9.8' and a bright magnitude 7.1, this globular just at the limit of resolvability presents us with a mottled circle of stars. There is no distinct nucleus here, just a gradual increase of stars toward the center.

From a pair of globular clusters we go to a pair of widely separated planetary nebulae. In the northwest corner of Lupus is NGC6026, just under 4 degrees north of η Lupi. At between 150x and 200x magnification, this faint (magnitude 12.5) nebula shows a fairly large disk (50") and a central star. (RA 16h 00m 1.4 s Dec -34° 32'.)

As a side trip before visiting the other planetary nebula, we can look to a double star close by (just over 1 degree to the northwest). Xi (ξ) Lupi is an easy double for both large and small scopes. Here is a pair of white stars of spectral type A separated by 10.5", and each one is magnitude 5.5.

Now at the far western edge of Lupus is IC4406, our other planetary nebula that lurks among some scattered galaxies. This is also called the Retina Nebula. High power should be used on this small, elongated nebula of magnitude 11 since its brightest, observable portion is almost stellar in appearance. The combination of smaller size and brighter magnitude means that this nebula is easier to spot than the other one, but high magnification is required to see it as an extended object. You can find IC4406 4.7 degrees northwest of α Lupi, or RA 14h 22.4m Dec -44° 09'.

There is a small collection of galaxies in this same area of Lupus. The brightest is NGC5643, a round spiral galaxy measuring 4.5' x 4.0'. If you have found our preceding planetary nebula, IC4406, an easy way to find NGC5643 is to move 1.8 degrees due east (or RA 14h 32.7m Dec -44° 10.4'). Otherwise move 2 degrees SSW of η Centauri, a 2nd magnitude star forming the northwest point of a triangle with α and β Lupi.

A couple of open clusters deserve mention, but are almost impossible to view due to their far southern location. These two clusters are situated near -55 degrees Declination, so you might want to keep these in mind if you ever find yourself observing near the southern hemisphere. NGC5822 is a very large cluster covering 40', or almost half again larger than the width of the moon. It sits well within the southern summer Milky Way, which is a rich star field to begin with. This cluster is a collection of 120 stars of magnitudes 9 to 12, making a combined brightness of magnitude 7.0. Its location is 2.5 degrees SSW of Zeta (ζ) Lupi, or RA 15h 5.2m Dec -54° 21'.

NGC5749 is a diminutive cluster by comparison measuring 8' across, of magnitude 9, and containing stars between 10th and 11th magnitude. Move 2.3 degrees west of NGC5822 to find this one (or center your scope at RA 14h 38.9m Dec -54° 31').

Lupus can be a challenge to observe due to its southern declination, and it takes patience waiting for some of its objects to get high enough to easily view. But it is well worth your time to view the interesting objects it contains.

Lynx (links)
Abbreviation: Lyn
The Lynx
Best observed between January 15th and April 15th

The American Heritage Dictionary defines a lynx as "a wildcat with thick, soft fur, a short tail, and tufted ears," and this is the animal Johannes Hevelius chose in 1687 for a constellation to fill in an area between the front of Ursa Major and the adjacent constellations of Gemini and Auriga farther west. It's not so much the pattern of stars that brought a lynx to his thoughts, but their dimness. The brightest star of Lynx is magnitude 3.1, and most others are dimmer than 4. Mr. Hevelius remarked that one would need the visual sensitivity of this nocturnal cat to observe this constellation.

To find Lynx, you can start at the Big Dipper. Follow the body of Ursa Major to its snout, and then look within the triangle formed by the bear's nose and the bright stars of Capella in Auriga, and the twin stars of Castor and Pollux in Gemini. The stars of Lynx are sprinkled here in a line going toward the southeast to just above the head of Leo the Lion.

There are few deep sky objects in Lynx that are comparatively bright, and without any conspicuous guideposts to point the way, some can be difficult to hunt down. However, the best ones are well worth the effort.

All by itself (both in terms of its place in our celestial sky and also its actual location in space) is NGC2419. This globular cluster has been nicknamed the "Intergalactic Wanderer," since it is far removed from the group of other globulars huddling in the Milky Way's halo. While a typical globular might be between 20,000 and 50,000 light years away, this recluse keeps its distance at approximately 290,000 light years. Despite its remoteness, it can be viewed fairly easily in amateur telescopes. It glows at magnitude 10.4 and measures 4.1' in diameter. A 5 inch scope will just bring out the round smudge, and while 8 to 10 inch instruments will make it more obvious, don't expect to see any individual stars. The brightest ones can't manage more than a 17th magnitude glow. But, for it to be seen this well overall must mean that it is quite a large and bright cluster, and would be quite an impressive sight if it were closer to us. NGC2419 is at RA 7h 38m 6s Dec +38° 53'.

Lynx's brightest galaxy, NGC2683, is an edge-on spiral of magnitude 10.4 and a size of 9.3' x 2.2'. It has a bright core and very dim, hazy outskirts. You'll see a sharp, thin oval contrasting very well with the background space. Point to RA 8h 52m 41s Dec +33° 25' 03", or start at Alpha (α) Lyncis, move 1.2 degrees southwest to a 7th magnitude star, then just over 5 degrees due west of here to this fine galaxy.

NGC2537 is a different kind of galaxy. At first glance it may look like a globular cluster. Its round shape of magnitude 12.1 is 1.7' x 1.5' across, and various toe-like extensions on one side has led to the nickname of the "bear claw galaxy." Look to RA 8h 13m 15.1s Dec +45° 59' 29", or 3.3 degrees NNW of 31 Lyncis.

A great double star is found in 19 Lyncis. It's in the northern half of the constellation at RA 7h 22m 52s Dec +55° 16' 53". Here is a primary star of magnitude 5.6 with a secondary of magnitude 6.5. They have a separation of 14.8" and a PA of 315 degrees. Any telescope can find and separate this double, but if you have the aperture and dark skies, you can see two other stars that are a part of the same system. Turning to a PA of 3 degrees and 4' away is a magnitude 8.9 star, and 72" from the primary at a PA of 287 degrees is a faint 11th magnitude star completing this quadruple system.

Lyra (ligh' ruh; also lee' ruh)
Abbreviation: Lyr
The Harp
Best observed between June 1st and September 1st

This is a small constellation, but due to its bright star Vega, it has attracted the attention of many different civilizations over the millennia. As a result, it carries many different mythical stories as well. Originally, those ancients inhabiting the Middle East saw the shape of a vulture in the stars of Lyra. The Greeks came up with the idea that the stars represented a kithara, or what we would call a harp. Over the centuries, these two concepts were combined so that some artwork shows the harp held by a vulture.

The Greek god, Hermes, is given credit for inventing the kithara. He found an empty tortoise shell, set some strings across it, and discovered that it made beautiful music. Over time, the kithara passed from Hermes hands to Apollo's, and finally to Apollo's son, Orpheus. It was Orpheus whose talent allowed him to play the instrument the most skillfully. The wild beasts were mesmerized by the music, and even the stones of the Earth stopped to listen. When Orpheus died, Zeus placed the kithara in the sky.

Asian cultures saw Vega as a princess who fell in love with and married a shepherd, represented by Altair, the brightest star in the nearby constellation Aquila. They were so enamored with each other that they neglected their earthly duties, and the princess's father banished them to the heavens to spend the rest of time separated by the great celestial river, the Milky Way.

Vega, which means "vulture" in Arabic, is the first of the three stars forming the "Summer Triangle" to rise during the early evenings of April and May. The other two are Deneb in Cygnus, and Altair in Aquila. At 9pm during the middle of August, Vega and its constellation Lyra are straight overhead. Vega is a blue-white star of spectral type A0 and is a mere 26 light years from Earth.

Near Vega (a bit more than 1.5 degrees to the northeast) is an intriguing object, Epsilon (ε) Lyra. This is a double star that is commonly called the double-double. Viewing it at low power will show you a typical double star separated by a very wide 209", but high power will show you why it has this unique name. Each component of this wide double is, itself, a double star, separated by about 2.7" and with position angles turned roughly 90 degrees with respect to the other.

Between Beta (β) and Gamma (γ) Lyrae is the most famous object in Lyra: M57, or the ring nebula. This is one of the finest examples of planetary nebulae in our sky. It is visible in any size scope due to its strong surface brightness, and its ring shape is easily recognizable. However, it's not truly a ring. Instead, we perceive a ring structure because of the thicker expanse of material along the periphery of the gaseous bubble. It is slowly expanding from the central star which, at close to magnitude 15, is invisible in scopes smaller than about 18 inches. The brightness of M57 is 9.0, and its diameter spans 2.5'. Look not quite half way from β to γ Lyrae, or RA 18h 53.5m Dec +33° 02', for this striking nebula.

Just under 1 degree east southeast of Theta (θ) Lyrae is NGC6791, a wonderfully rich galactic star cluster, 16' across and of magnitude 9.5. Since it sits in a rich part of the Milky Way, it is difficult to recognize the boundaries of this star cluster. There is a gradual increase in stellar density toward the center of the cluster of several hundred faint stars. Of course, the larger the telescope you have, the more stars you will pick up, and thus the more impressive this cluster will be. You can find NGC6791 at RA 19h 20.7m Dec +37° 51'.

All by itself in the southeast part of the constellation is M56, a globular cluster weighing in at magnitude 8.3 and a size of 7.1'. This is a fairly rich, condensed globular with stars ranging in individual brightness from 11 to 14. If your telescope is accurately polar aligned, dial in this cluster at RA 19h 16.6m Dec +30° 11', or start at magnitude 4.3 Theta (θ) Lyrae and move directly south 8 degrees. (The difference in RA of θ and M56 is only 22 seconds, so it's almost directly due south.)

Microscopium (my' kroh skoh' pee um)
Abbreviation: Mic
The Microscope
Best observed between August 1st and September 1st

This is a more modern constellation created by Nicolas-Louis de Lacaille in or about the year 1750. He named it along with Telescopium (found farther to the south) to honor two instruments which revolutionized the study of science. To the ancient observers, the area occupied by Microscopium had no named figures and was known as a vast area called "the sea" rising in the sky above the horizon. Microscopium is made of very dim stars (4.7 and dimmer) south of Capricornus and between Sagittarius and Grus. It passes through the meridian at around 11pm during the second half of August.

Making your way around Microscopium is difficult because there are no bright stars to use while navigating through this area. One of the brighter stars of Microscopium, Alpha (α) Microscopii, is an easy double star within the grasp of any telescope, and provides a good starting point. Individually, they shine with magnitudes 5 and 9.8, giving a naked eye integrated brightness of 4.9. Twenty arc seconds separate the two, so low power is all that's necessary to split them. These two stars are very similar to the sun, having a spectral type of G8, compared to G2 for the sun.

There are a number of galaxies in Microscopium, but all are faint. One of the brightest that we find here is a fine example of a spiral galaxy: NGC6925. We can star hop from Alpha by going 2 degrees west to a star of 5.5 magnitude, and then another two degrees northwest to NGC6925 (or go to RA 20h 34m 21s Dec -31° 58' 50"). This magnitude 12.1 galaxy is turned very close to edge on and measures 4.4' x 1.2', so you'll see a dim spear of light oriented almost straight north-south. In very large instruments or CCD images, some texture can be seen along the tightly packed spiral arms.

Monoceros (mon oh' sair ohs)
Abbreviation: Mon
The Unicorn
Best observed between January 1st and March 15th

If you browse through the winter skies with the naked eye, you will enjoy some of the best sights available of the heaven's brightest stars and most recognizable constellations. But then you will come to a rather lackluster region east of Orion. At first glance, this area may look uninteresting, but it compensates for its lack of bright stars with some excellent deep sky objects. The winter Milky Way cuts right through this area occupied by the constellation of Monoceros, so it is rich in nebulae and star clusters.

It was in the early 1600's that German astronomer, Jakob Bartsch, established the constellation of Monoceros out of some dim stars stretching between Orion and his two hunting dogs, Canis Major and Canis Minor. The Latin name Monoceros is derived from the Greek word "monokeras," or "one horned," so this constellation represents the mythical creature called a unicorn. It might, however, require a fantastic imagination to see a unicorn in this pattern of stars.

One of the brightest stars of this constellation is Beta (β) Monocerotis, among the finest multiple stars visible through small telescopes. When Sir William Herschel discovered this system in 1781, he described it as a "beautiful sight." Having a combined magnitude of 3.8, it is a triple system of stars that are uniquely very similar in both brightness and spectral type. β Mon A, magnitude 4.7, and β Mon B, magnitude 5.2, are separated by 7.4", while β Mon C is of magnitude 5.6 and between the two about 2.8" from the B component. All three are of spectral type B2 to B3 and appear as a trio of pure white stars forming a very flat triangle. Another notable binary star is Epsilon (ε) Monocerotis. Its components are of magnitudes 4.3 (spectral type A5) and 6.7 (spectral type F4) and are separated by 13".

Another binary star worth mentioning is Plaskett's star. It is noteworthy because it is one of the most massive pairs known. Located at RA 06h 37.6m Dec +06° 11', or about 1.5 degrees to the southeast of 13 Monocerotis, this binary was first studied by J. S. Plaskett in 1922 and is made up of two stars of spectral type O separated by a true distance of only about 50 million miles. It is estimated that the two stars have masses of 40 and 60 suns and shine with an apparent magnitude of 6.06.

Among the deep sky objects that Monoceros presents to us is M50, a medium-sized open cluster that is easily within the grasp of binoculars. Located approximately a third of the distance along a line connecting Sirius and Procyon and shining at magnitude 7.2, its coordinates are RA 7h 03' Dec -8° 21'. The main group in this cluster is roughly 10' in size. Long exposure photographs bring out about 200 stars within this cluster.

In the northern regions of the constellation are two more splendid star clusters and associated nebulosity. Starting with NGC2244, this is a bright cluster in the shape of a bent rectangle sitting in the center of a vast, doughnut shaped nebulosity known as the Rosette Nebula. The brightest sections of the nebula, which as a whole occupy an incredible 80' of the sky, actually are given separate NGC designations of 2237, 2238, 2239, and 2246. With dark skies, a 6" telescope will begin to show the soft glow of this dim nebula, and you may not be able to view the entire structure at once depending on your magnification. This complex of stars and nebulosity is centered at RA 6h 30.3m Dec +5° 03'.

Farther north is NGC2264. The nickname of this cluster is the "Christmas Tree" cluster because of the way it resembles the outline of a Christmas tree. It is a large cluster of very young stars about the same size as the lunar diameter, so low power is required to fully see it. Studies suggest that some of the members of the cluster are still undergoing gravitational collapse on their way to main sequence status. The brightest star of the cluster, 4.6 magnitude S Monocerotis, forms the trunk of the tree and shines with the light of 8500 suns. Long exposure images show strong nebulosity near S Monocerotis, but the most recognizable feature is found just off the point of the tree, which points south, and is known as the Cone (or Conus) Nebula. This is a dark nebula about six light years in length, or spanning about 10 arc minutes in the sky, contrasted by brighter clouds surrounding it. You can find NGC2264 at RA 6h 41m Dec +9° 48'.

A much easier nebula to spot is Hubble's variable nebula, or NGC2261. This is located to the southwest of NGC2264 at RA 6h 39.2' Dec 8° 44'. It appears as a fan-shaped nebula about 2 to 3 arc minutes in size and of fairly high surface brightness, which Edwin Hubble found to vary in overall magnitude, size, and structure (sometimes over the short span of a few months). The nebula shrouds the variable star R Monocerotis, but the variability of this star does not seem to coincide with the behavior of the nebula.

Ophiuchus (oh fee yoo' kuss)
Abbreviation: Oph
The Serpent Bearer
Best observed between June 1st and July 31st

This constellation, rich in history, is formed by stars of medium brightness. It lies north of Scorpius and reaches highest elevation between 11pm and midnight during the middle of June. Its name comes from a Greek word meaning "serpent bearer," and goes back about 4,000 years. Traditionally, Ophiuchus represents Asclepius, considered to be the god of medicine and son of Apollo and Coronis. When Asclepius happened to kill a snake, another snake came along and brought the first one back to life with herbs, and that is how Asclepius first learned about medicine. He became a physician and eventually honed his trade so well that he began to adopt the serpent's talent for bringing the dead back to life. Hades, ruler of the underworld, was worried by this since it threatened the arrival of new souls into his domain, so he struck a deal with Zeus to kill Asclepius. Since the physician was so accomplished in the science of medicine, Zeus honored him by placing him in the sky with a snake stretched across his body. Thus we see Ophiuchus standing between the two halves of the snake, Serpens Cauda (on his east side) and Serpens Caput (toward the west). The symbol of the American Medical Association, a snake coiling around a staff, takes the inspiration for its design from this story. It's interesting to note that the sun spends more time in Ophiuchus than it does in Scorpius, so instead of the scorpion, Ophiuchus could easily have been the next sign of the zodiac after Libra.

Ophiuchus stands on the western edge of the summer Milky Way, so you may choose to start your observations with binoculars. His eastern leg is immersed in the rich star fields next to Sagittarius with his torso stretching northward. A dark nebula worth trying to spot (most easily on film with a standard lens and long exposure) is the Pipe Nebula. It stretches 7 degrees from one end to the other in the area southeast of Theta (θ) Ophiuchi.

Like Sagittarius, Ophiuchus is dotted with globular clusters which are the main draw to the constellation (see the list below). M12 is an easy target to find toward the west side of the constellation. It is a fairly large (12' diameter) globular lacking a well-defined core and showing a clumpy appearance. M10 is a short distance southeast of M12, is of similar size, and its distinct stars are easily resolved in medium-sized scopes. Another rich cluster is M14, a pleasing sight through a telescope of any size. Larger instruments will begin to resolve hundreds of individual stars.

Going down the size scale are other clusters of a more difficult nature to find. M9 and M19 are embedded in the Milky Way star clouds. Both are rather small (about 6' diameter) but possess their own unique characteristics. M9 shows a dense core glowing with a soft intensity, and M19 (the smallest of the group) has an odd oval appearance.

M62 and M107, near the south and west edges of the constellation respectively, are also small and shining with a subtle light. M62 is a little more interesting due to its brighter core. M107 shows little more than a very small, soft glow.

Object	Coordinates (RA and Dec)		Magnitude
M12	16h 47.2m	-1° 57'	8.0
M10	16h 57.2m	-4° 06'	7.6
M14	17h 37.6m	-3° 15'	9.4
M9	17h 19.2m	-18° 31'	8.9
M19	17h 02.6m	-26° 16'	8.3
M62	17h 01.2m	-30° 07'	8.2
M107	16h 32.5m	-13° 03'	10.1

British astronomer Edmund Halley determined that stars exhibit proper motion by comparing his observations to those made long ago. Ophiuchus contains Barnard's Star, the second nearest star to Earth and the one with the largest proper motion in the sky. It's a magnitude 9.53 red dwarf star (spectral type M5), and each year this runaway star moves 10.29" practically straight north. Careful observations could reveal its motion to you after only a few years. Refer to the finder chart included near the front of this book. The chart's inset spans about 22' at a magnification of 100x and shows Barnard's Star by my observation on October 4, 2002. Its movement is toward the top. To give you an idea of its speed in a telescopic view such as this, in the year 1960 it was in the middle just off the tip of the arrowhead pattern of stars. By about 2035, it will have exited the field of view. That's equal to about one-third of our moon's diameter over the course of those 75 years.

Orion (oh rye' un)
Abbreviation: Ori
The Hunter
Best observed between December 15th and March 1st

We could call Orion the heavyweight among all the constellations. It is probably the most well-known group of stars with, perhaps, the exception of the part of Ursa Major called the Big Dipper. On the celestial equator is a line of three stars forming the very recognizable "belt" of Orion. The faint stars and nebulosity below this represents his "sword", and other faint stars on each side of Orion's body show one arm raising a club and the other holding a shield against Taurus, the bull charging out of the northwest.

Mythology portrays Orion as a strong hunter. He claimed that no living thing could kill him, so to prove him wrong, the goddess Hera sent a scorpion to attack him. After falling victim to the scorpion's sting, Orion squashed the pest, but soon died from the venom. We now see this relationship between the scorpion (the constellation Scorpius) and Orion on fall and spring nights. The two characters are placed on opposite sides of the sky, so when one rises, the other is defeated and sets below the horizon.

You can begin observing Orion before it is fully dark. Use high power to see Rigel (Beta (β) Orionis), the star marking Orion's western foot, and its magnitude 6.7 companion 9" away and holding a PA of 202 degrees. It is easier to view this double star while the sky is still in twilight because of Rigel's overpowering brightness. Diagonally across the body of Orion from Rigel is Betelgeuse (Alpha (α) Orionis). This is an M2 supergiant of a striking orange color. Betelgeuse has become a red giant star since it has exhausted most of the hydrogen in its core and is now fusing helium to remain shining. For this reason, it puts out an enormous amount of energy, but only a fraction of it is in visible light. The result is that its outer atmosphere has inflated to make the star a truly bloated behemoth. Estimates of its size put its diameter matching the orbit of Mars or more, which also creates a stellar density of an exceedingly small nature. It's millions of times less than the sun. The three stars forming Orion's belt are all hot O or B type stars and in the neighborhood of 1,300 to 1,500 light years away. All three produce many thousands of times the light of the sun, which is why they are so obvious to our eyes from such a distance.

The best deep sky showpiece in Orion, and one of the best in the northern night sky, is the Orion Nebula. "...I felt that you could not avoid casting your eyes upward to the great nebula in Orion, and I certainly expected that you would do so," spoke C. Auguste Dupin to his companion as they walked down a darkened Parisian street in Edgar Allen Poe's, "The Murders in the Rue Morgue." I often find the same great nebula drawing my own attention on winter nights as it has undoubtedly done to countless others before and after this 19th century author. It is seen in dark skies with the unaided eye as a distinct fuzzy glow forming part of the sword beneath Orion's belt. This book's cover photo shows Orion with the pinkish glow of the nebula just below and to the right of center. Binoculars capture a wispy fog surrounding its central stars, 3-inch telescopes will bring out a fan shaped nebula, and 8-inch and larger scopes will show detailed knots and tendrils extending away from the main glow. The entire complex of gas is made to fluoresce by Theta1 (θ1) Orionis, called the Trapezium. Most views of this group of stars will show 4 members of magnitude 5.4 to 6.7 about 30" in expanse, but high power and large aperture will bring the total to 6. Adjacent to the main mass of M42 is M43, a roughly comma shaped object separated from M42 by a dark rift. To the north is a dusty reflection nebula with the designations of NGC1975, 1973, and 1977. It's curiously named the "Running Man Nebula" from the shape that is traced out by the dark lanes that separate distinct sections of the nebula. 10-inch apertures and larger are best for the dim glow of the "Running Man."

Another well-known nebulous region surrounds Zeta (ζ) Orionis. This one, however, is notorious for its difficulty to observe. NGC2024, the "flame" or "flaming tree" nebula, is not too elusive about 15' east of ζ. Seen with a 6-inch or larger telescope, it's a roughly circular emission nebula with branching dark lanes crossing it toward the north. Below this is the famous "Horsehead Nebula." The glowing region, IC434, is a 1 degree long spike of nebulosity pointing south from ζ. Intruding into this is the dark nebula B33, forming the shape of the horse's head.

While some claim to have seen it on dark nights with plenty of aperture (I claim to have seen it only once through an 11-inch telescope on a night of exceptional darkness and clarity, and I still question whether I was actually seeing it, or if it was just in my imagination), most will only capture it on film or CCD images.

2.5 degrees northeast of ζ is M78, one of the more challenging Messier objects to find, but unmistakable when you do. The first time hunting this nebula for me took close to 15 minutes, and I probably passed over it a few times before knowing that I had it in my eyepiece. It will be easy to overlook, like I did, if using low power (since the brightest portion of it is fairly small, and small telescopes will allow it to masquerade as simply a double star that sits at the nebula's center). It is cometary in appearance with a dusty glow surrounding the magnitude 8.2 double star. The nebula shines at magnitude 8. Hunt down M78 at RA 5h 46m 42s Dec +00° 3'.

With all that Orion has to offer, it is easy to brave the cold temperatures of winter to spend some time with his treasures.

Pegasus (peg' uh suss)
Abbreviation: Peg
The Winged Horse
Best observed between September 1st and November 15th

Mythology tells us that the exotic creature called Pegasus was born from the events related to the battle that Perseus had with Medusa, a Gorgon whose appearance was so ugly that any mortal who gazed upon her would turn to stone. To avoid that end, Perseus used the reflections in his shield to locate Medusa, and then beheaded her with his sword. He carried her head with him to the seashore, and as drops of Medusa's blood fell into the breaking waves, Pegasus took shape from the mixture and soared skyward. The winged horse is always depicted as white, the same color as the sea foam from which he emerged.

As a very much admired horse, he was captured by Athena, who took him to Mount Helicon, where he kicked the ground and started the sacred spring of Hippocrene. In other times, he was the companion of Bellerophon, a youth who (with Pegasus' help) destroyed Chimera, a greatly feared monster that breathed fire and was part lion, part goat, and part dragon. To honor the beloved horse, the gods gave him an eternal place in the sky, very near the zenith at 1am on September 1st (8pm at the end of October). The most recognizable feature of the constellation is the pattern of four prominent stars forming the "Great Square" of Pegasus (if we borrow the Alpha star from Andromeda) southwest of

Cassiopeia and north of Aquarius.

The Arabic names given to many of the stars in Pegasus describe his body parts or are a tribute to the reverence shown toward the horse. Markab (Alpha (α) Pegasi) may mean "horse's shoulder" or "saddle," Scheat (Beta (β)) translates to "shin" or "foreleg," Algenib (Gamma (γ)) is the "side," Enif (Epsilon (ε)) marks the "nose," Homam (Zeta (ζ)) is thought to mean "hero," and Sadalbari (Mu (μ)) most likely translates to "excelling one."

One of the sky's best globular clusters is located at RA 21h 30.0m Dec +12° 10' or about 4 degrees northwest of Enif, Pegasus' nose. This is M15, which glows at magnitude 6.5. The most striking feature of this cluster is the bright, compact core that blazes with an almost stellar brilliance. Overall, the size of this cluster (seen as either a fuzzy circle in binoculars or a sprinkling of thousands of stars in large telescopes) is about 10' in diameter. Another uniqueness of this cluster is the presence of k648, a small planetary nebula which seems to be a true member of the cluster and not a foreground object. Don't be disappointed if it's not visible, however. Its magnitude is 13.8, and it measures only 1" across. That's not surprising for an object such as this from 35,000 light years away.

NGC7331 is an easy NGC object to spot near the northern boundary of Pegasus. It is a galaxy with a glow of magnitude 10.4 and measures 10' x 2.5' centered at RA 22h 37.1m Dec +34° 26', conveniently aligned with Mu (μ) and Eta (η) Pegasi when drawing a line northward. The best description of NGC7331 would be to say that it very closely resembles M31, the great Andromeda Galaxy. NGC7331 is almost 25 times more distant, so it appears correspondingly smaller. Through a telescope it shows its oval structure that is tilted only about 20 degrees from edge-on.

Not too far away, at RA 22h 08m Dec 31° 21.5', is another spiral galaxy of magnitude 11.3. In contrast to NGC7331, this galaxy is turned nearly face-on. It measures a small 2.5' in diameter, but it has fairly bright and tightly wound spiral arms, which makes an overall structure that increases in brightness gradually from its outskirts to the nucleus. If you find NGC4217 by starting at the nearby star Pi (π) Pegasi, simply move 2 degrees south and 2' in RA west.

The "Great Square" of Pegasus is easy to find and one of the hallmarks of the fall sky. The full extent of the constellation gives us several appealing objects to study.

Perseus (per' see uss)
Abbreviation: Per
The Hero
Best observed between November 1st and February 28th

The character that we call Perseus interacted with many others in mythology. He was the son of Zeus and the mortal Danae. His most storied accomplishments begin with his intrusion into the cave of the Gorgons, a group of sisters so hideously ugly that any mortal who had the unfortunate chance of setting eyes on one of them would instantly turn into stone. He hunted down Medusa, the only mortal Gorgon, by safely viewing the reflections in his shield. With his sword, Perseus successfully severed Medusa's head to at least end her particular threat to the surrounding population.

In his travels immediately following his encounter with Medusa (who's head he still had with him), he came upon a princess named Andromeda, who was chained to the rocks at the edge of the ocean. The story behind this situation comes from the fact that Andromeda's mother, Cassiopeia, was incredibly boastful, and thus was disfavored by the gods. A sea monster, Cetus, was sent to destroy the coast of the land ruled by Cassiopeia and her husband, Cepheus, as punishment for her egotism, and they sent their daughter as a sacrifice to send away the monster. Just before Cetus reached the panicked princess, Perseus appeared. He held up the head of Medusa, allowing Cetus to get an eyeful of the terrible face, and the monster (at this point an immense stone slab) fell harmlessly into the sea. An inadvertent outcome of this episode was the creation of the winged horse Pegasus. While Perseus was handling Medusa's head, drops of her blood dripped into the sea foam at the edge of the water and out of this mixture appeared Pegasus.

You should have recognized many familiar names here. All but Medusa are represented in the sky with constellations of their own, and in addition, are all in the same half of the sky between 20 hours and 3 hours Right Ascension. Perseus is found between the pentagon-shaped constellation of Auriga and the W shape of Cassiopeia.

Perseus stands at a high declination, closer to the celestial pole than to its equator. During late fall it actually rises near midday, crosses the meridian around 10pm, and sets close to dawn. It straddles the winter Milky Way, so we can presume that the constellation provides a good collection of objects to observe.

A star of special significance is in the southwest part of the constellation. Its name, Algol, is taken from the Arabic Al Ra's al Ghul, or "the demon's head." Many historical cultures have considered this star to be evil, or at least to bring terrible luck. The simple reason for these feelings is the fact that Algol is an eclipsing binary star. Therefore, its light varies, shining most of the time at magnitude 2.1, but then dimming briefly to 3.4 over a period of 2 days, 20 hours, 48 minutes, and 56 seconds. Indeed, observing a naked-eye star behaving this way when all the others shine with steady light might suggest to those in ancient times that something supernatural was involved. However, the explanation is that there are two stars in a close orbit which brings a much dimmer G5 star in front of its brighter B8 companion at regular intervals, thus resulting in a drop in the combined light of the system lasting about ten hours. It's a simple matter to gauge Algol's brightness by comparing it to Rho (ρ) Persei, which itself is another variable star, but shines with a brightness of magnitude 3.3 most of the time. Rho is a little over 2 degrees away.

The prize for the most photogenic object in Perseus is probably the twin star clusters of NGC869 and NGC884. The common name for the pair is, appropriately, the Double Cluster. Each share similar brightness (magnitude 4.4 for for NGC884, and 4.3 for NGC869) and size (30'). A total of about 150 stars populate 884, and 200 for 869. Most of the stars in each cluster have a white or bluish-white tint, but a few stand out with more of an orange hue. You'll easily see the pair forming a fuzzy area to the naked eye in dark skies in the northwest corner of the constellation. Use low power through the telescope to fit both clusters in the same field of view for the best effect. Look closely at NGC869 and you may see, with a little imagination, the figure traced by stars of what some people see as a gingerbread man.

Not too far away is the planetary nebula M76, or the "Little Dumbbell". Its name comes from the fact that indeed it does resemble M27, the Dumbbell nebula in Vulpecula. This one, however, measures only 65" and shines with a dim magnitude 12 glow. Its brightest part forms a rectangular shape in a telescopic view, but there are also looping filaments of gas along its long axis mostly invisible to the eye but apparent in deep CCD or film exposures. M76 is off by itself one degree north-northwest of Phi (φ) Persei or, more precisely, RA 1h 42m 18s Dec +51° 35'. It is definitely one of the more challenging Messier objects.

M34 is a collection of about sixty stars covering 35', or about the same size as the full moon. It's overall magnitude is 5.2, and this makes it easy for binoculars or the finder scope about 5 degrees northwest of Algol.

NGC1023 is our only bright galaxy in Perseus. It's an oval 8.7' x 3.3' in expanse with a magnitude of 9.5. Its appearance is reminiscent of M31, the famous Andromeda Galaxy, but, of course, many, many times smaller as viewed from Earth. NGC1023 is 3.7 degrees south of M34, or RA 2h 40m 24.1s Dec +39° 03' 46".

If you are up for a real challenge, try NGC1499. This, the

California Nebula, requires a large scope, very dark skies, and perhaps even a nebula filter. Its dim light is spread over a truly huge 145' x 40'. Look to Menkib, or Xi (ξ) Persei, which is the star that illuminates NGC1499, and move north about 3/4 of a degree.

Phoenix (fee' nicks)
Abbreviation: Phe
The Phoenix
Best observed between October 15th and November 30th

The constellation that we know as Phoenix evolved from what the ancient Arabs called Al Zaurak (the Boat), as well as Al Rial (the Ostriches). Johann Bayer first identified this constellation as Phoenix on his star atlas of 1603. The Phoenix is a mythological bird which, at the end of its long 500 year life span, would build a fire and throw itself upon it. Then, from the ashes, a new Phoenix would rise to begin a new life cycle. Its red and gold plumage is reminiscent of the fire that both ends and begins its life.

Phoenix is in a part of the sky that hugs the southern horizon from the southern United States as it crosses the meridian. To find Phoenix, start with the easily recognizable constellation of Cetus (the whale, or sea monster), go south through the dim stars of Sculptor, and just above the horizon you'll see the 2nd magnitude and fainter stars of Phoenix. Its brightest star, Alpha (α) Phoenicis, forms a rough equilateral triangle with Beta (β) Ceti and Alpha Piscis Austrini (Fomalhaut).

There is a large collection of galaxies in Phoenix, but many are dim and too close to the horizon to be worth finding. However, just a bit more than 2 degrees northeast of Gamma (γ) Phoenicis is a moderately bright spiral galaxy of magnitude 11.6. It's tilted between face-on and edge-on, so it presents itself as a more or less featureless oval, measuring 5.7' x 1.9', and oriented almost perfectly east to west in its long axis. NGC625 can be dialed up at RA 1h 35m 5s Dec -41° 26' 11".

The preceding galaxy can be glimpsed in most backyard telescopes in clear, dark skies. Another one would be IC5325, a slightly fainter magnitude 12.0 spiral at the western edge of Phoenix at RA 23h 28m 43s Dec -41° 19' 58", or 1.7 degrees northwest of Iota (ι) Phoenicis. This one shows its face to us, but is rather small at only 2.7' x 2.5'.

The remaining galaxies in Phoenix are much more challenging, but those of you with large aperture telescopes might want to look in the area of Beta (β) Phoenicis. In a two degree diameter circle around this 3.3 magnitude star are four faint galaxies. Starting 1 degree northeast of

β, you'll see IC1633, a magnitude 12.2 elliptical galaxy 2.9' x 2.4' in size (RA 1h 9m 55s Dec -45° 55' 56"). Moving half a degree west from this galaxy will bring you to PGC03939, a magnitude 13 lenticular galaxy measuring a tiny 1.1' x 0.9' (RA 1h 6m 49s Dec -45° 49' 29"). Go half a degree southwest to find PGC03847, another lenticular galaxy of magnitude 13 and 1.3' x 0.7' (RA 1h 5m 00s Dec -46° 4' 35"). Finally, go south 1 degree from here (or half a degree southwest of β Phoenicis) and you might catch a glimpse of PGC03845, an elliptical galaxy 1.1' x 0.7' in size and of magnitude 13, as well (RA 1h 4m 54s Dec -46° 59' 59").

Going for an easier target might be just what we need now. Look again to the western side of the constellation for Theta (θ) Phoenicis. This is a double star of magnitude 6. Go to high power to split this pair of magnitude 6.5 and 7 stars that are 4" apart.

Because Phoenix is so close to the horizon, it takes patience and determination to observe most of what it has to offer, but to fully explore the sky, it's worth pointing your telescope down to it.

Pisces (pie' seez)
Abbreviation: Psc
The Fish
Best observed between October 1st and December 30th

Although this constellation is one of the prominent ones of the zodiac, it can be easily overlooked because it has no star brighter than 3rd magnitude. It represents two fish: one beneath Andromeda, and the other below the "Great Square" of Pegasus. They are joined to each other at Alpha (α) Piscium and thus form a skinny "V" pattern pointing toward the southeast.

The ancient Babylonians, Persians, Romans, Greeks, and Turks all saw fish in this group of stars. One story depicts Aphrodite and her son Eros who jumped into a river to escape a giant named Typhon. They turned into fish and swam away. So that they wouldn't be separated, they tied their tails together with a cord. In fact, the Alpha star mentioned above is named Alrisha, Arabic for "knot" or "cord."

South of Pegasus's square is the circlet of stars marking the head of the western fish. Seven stars make up the circlet, and the one which stands out from the others is the variable star, TX Piscium (also designated 19 Piscium), the eastern most star of this group. It varies slightly from magnitude 5.5 to 6.0, but one look at it will show that what makes it special is its color. Although it doesn't have a catchy nickname like the Crimson Star in Lepus or the Garnet Star in Cepheus, its red color is no less striking. Try using the different color sensitivities of your retina by looking directly at it, and then to the side. You should notice that its color is rich when in the center of your vision, and then fades to a lighter pink or almost white as you move your gaze away from it.

Pisces is home to a decent number of galaxies, but the only one big and bright enough for Charles Messier to add to his catalog is M74. Its magnitude 11.0 glow and 9' x 9' size can be spotted 5' in RA east and 0.5 degrees north of Eta (η) Piscium (or RA 1h 36.7m Dec +15° 47'). This face-on spiral should appear as a faint, circular glow that lacks much detail. In fact, early astronomers found it to be so elusive that after its discovery in 1780, there were differing opinions as to what type of object it really was. John Herschel decided that it was a globular cluster in his General Catalogue in 1864 when others thought it was a galaxy, and it wasn't until 1893 that the first detailed photograph showed its spiral arms.

Another spiral galaxy within range of amateur instruments is NGC488. It is almost as bright as M74 at magnitude 11.2, but much smaller, measuring roughly 3' x 3.5'. This is a compact spiral with tight arms that serve to more evenly distribute its light across its face. The following are a few ways to find NGC488: look about 2/3 the distance between Alpha (α) and Epsilon (ε) Piscium and one degree south of this line, or starting at ε, drop down 2.5 degrees and move east 19' in RA, or lastly, dial up coordinates RA 1h 21.8m Dec +5° 15.4'. Once you find it, go to a high power eyepiece to bring it into better view.

An easy double star to view in any telescope is Zeta (ζ) Piscium. This pair of magnitude 4.2 and 5.3 stars sit less than 1/4 degree south of the ecliptic and are separated by 23.5". They are of neighboring spectral types (A5 and F6), so the color contrast between the two may be too slight to detect. I see only shades of gray-white, but since individuals' eyes are different, some may see a subtle yellow or blue tint between them.

Piscis Austrinus (pigh' sis oss' trin uss)
Abbreviation: PsA
The Southern Fish
Best observed between September 1st and October 15th

This constellation is one of the more ancient ones, known to the Greeks and Romans as a single fish drinking from the water poured out of the jar held by Aquarius standing farther to the north.

The easy way to find Piscis Austrinus in September or early

October is to look to the south near the meridian between 9pm and 10pm. About 1/4 or 1/3 the distance from the horizon to the zenith is an area of the sky which is quite featureless to the naked eye. The exception to the blandness of this region is Fomalhaut, the constellation's only 1st magnitude star, and the only star of this brightness within the immediate area. The closest star of similar stature is Altair, lying 77 degrees to the northwest, and glowing at magnitude 0.8. Fomalhaut (from the Arabic "Fum al Hut," which means "mouth of the fish") is an A3 type star only 22 light years away. Sometimes called the "Solitary One," Fomalhaut signals the approach of cool autumn temperatures when seen during the mid-evening hours.

Oddly, Fomalhaut has been described by early astronomers as a red star. We've seen the same error made with Sirius, the brightest star in the constellation Canis Major (and also the star with the brightest apparent magnitude in the sky). These two stars are both at low declinations, so are frequently seen close to the horizon. It's possibly this fact that makes the light from these stars refract through the Earth's atmosphere in such a way that makes them sparkle with prismatic colors, mostly toward the red end of the spectrum.

Like all bright stars, Fomalhaut has held individual importance in many ancient civilizations. The Syrians saw it as a symbol of Dagon, their god of the sea. In ancient Persia, Fomalhaut, along with only three other stars (Regulus, Antares, and Aldebaran), was considered a "royal star" of heaven.

Piscis Austrinus has its fair share of galaxies, but they are all very faint and difficult for small telescopes. However, there are a couple we could hunt down. The brightest galaxy is NGC7314 (also known as ARP14), an 11.7 magnitude spiral at RA 22h 35.8m Dec -26° 03', or 1.5 degrees northwest of Epsilon (ε) Piscis Austrini. Its faint light is spread across 4.5' x 2.0'.

IC5271, 1.3 degrees south southeast of Delta (δ) Pisces Austrini, is a sharp-edged, 2.6' x 0.8' galaxy of magnitude 12.6, RA 22h 58m Dec -33° 44.5'. If at first you don't see it, try all the tricks for dim object observing: shielding your eyes from stray light with a hood or your hands, averted vision, or tapping the scope gently to induce a wobble in the image. Sometimes a moving object will snap into view.

An easier object is Beta (β) Picsis Austrini, a double star with a wide separation suitable for very low magnifications. β has a spectral type of A0, making it almost identical to Fomalhaut in color. Easy to find southwest of the bright star, β is widely separated with 30" between the 4.4 magnitude primary and the 7.9 magnitude secondary.

Puppis (pup' iss)
Abbreviation: Pup
The Poop Deck
Best observed between January 15th and February 28th

The southern sky holds the giant ship Argo, which is divided into several individual constellations. Puppis is the poop deck, or stern, of the ship oriented with its bow pointing south and its stern standing vertically. The stars of Puppis are 2nd magnitude and dimmer and sit east of a line drawn between two conspicuous winter stars, Sirius and Canopus (or substitute the horizon if Canopus is not visible below Sirius from your latitude). If you follow the line made by the three stars of Sirius, Delta (δ), and Eta (η) Canis Majoris and extend it southeast, your gaze will immediately begin traversing Puppis. The entire constellation slants from directly south of Canis Major to its east. You may notice that the Greek lettering of this constellation, the system that was established by the German astronomer Johann Bayer, is incomplete (there is no Alpha Puppis, for example). When Bayer distributed the letters for the stars in this region, he did so for the old, very large constellation of Argo Navis. Later, after this was broken up into the separate constellations of Puppis, Vela, and Carina, the Bayer designations for the stars were preserved as they still are today.

We have seen quite often that constellations stretching along the Milky Way are rich in star clusters, and Puppis is no different with its clusters outnumbering all other deep sky objects by a wide margin. Since Puppis is so close to the horizon, those objects that are near the northern areas of the constellation are the best targets. Two of these are M46 and M47. They can be mentioned together since they are only 1.3 degrees apart from each other. Even though they share the same neighborhood, they have much different personalities. M47's stars are a mixture of almost a dozen bright stars (the brightest being mag. 5.7) and another 30 or so dim stars that contribute to a total magnitude for the cluster of 4.4. M47 is located at RA 7h 36.6m Dec -14° 30'. M46 contrasts nicely, being a magnitude 6.1 cluster made of over one hundred noticeable stars of mostly magnitudes 9, 10, and dimmer. One interesting feature of this cluster is the magnitude 10.1 planetary nebula on the northern edge of it. This is NGC2438, a round halo measuring about 1' in diameter. Both M46 and M47 are close to 30' in size.

Stay in the neighborhood and you will see another cluster, NGC2423, about 0.6 degrees north of M47. This magnitude 6.7 cluster packs about 60 dim stars in an area 19' in diameter.

Moving south, we find M93. This cluster has a magnitude of 6.2, a size of 22', and a location of 7h 44.6m Dec -23° 52', or 1.5 degrees northwest of Xi (ξ) Puppis. It appears stretched along the southeast to northwest axis with lanes comparatively empty of stars, which give the impression of dividing the cluster into 4 sections.

As a change of pace, let's move southeast, past ξ Puppis, about the same distance as M93, but on the opposite side. Here is a challenging diffuse nebula designated NGC2467. It could be described as masquerading as a planetary nebula, since its overall shape is round with noticeable mottling to the edges. It even has an 8th magnitude star sitting in the middle of the nebulosity to add to the disguise. The size of this magnitude 7 nebula is respectable at 16', making for medium surface brightness. Crowding it is another star cluster, H2. This modest cluster has about 20 11th magnitude stars covering 9'.

A considerable distance away, near the western boundary of the constellation, is NGC2298. Fairly easy to find, but hard to resolve into stars, this is a globular cluster of magnitude 9.4 and measuring a small 6.8'. It is well down in the sky, so it's beneficial to wait until it's as close to the meridian as possible before observing it. You'll mostly see a imperfectly round smudge at RA 6h 49m Dec -36° 00'.

Fans of variable stars will want to seek L2. As a member of the long period variable class, this star ranges in brightness from approximately 3 down to less than 6, and during the process passes through the narrow range in spectral type from M5 to M6. It stays above the horizon only for a few hours, but that is not a problem since its period is a leisurely 141 days, and both sides of its light curve are equal. Even if you are not measuring its variability, you might still want to do an appraisal of its unique red color.

Pyxis (pick' siss)
Abbreviation: Pyx
The Compass
Best observed between February 15th and March 15th

The giant constellation of the ancient ship Argo Navis is, in modern times, broken up into the 4 distinct, smaller constellations of Vela (the sail), Carina (the keel), Puppis (the poop deck), and this constellation of Pyxis (the compass), which came along after the original three were established. The French astronomer, Nicolas-Louis de Lacaille, chose this navigational device to name the group of stars east of Puppis and Canis Major, even though the mythical ship Argo sailed before the invention of the compass. Composed of magnitude 3.5 and dimmer stars, it takes some effort to point out Pyxis. If you make a line between the two bottom bright stars of Canis Major (Epsilon [ε] and Eta [η]) and extend that line directly east about 4 to 5 times the length of that line (passing through Puppis), you'll arrive at Pyxis. Its three brightest stars form a shapeless line running roughly north and south. Pyxis is on the meridian at 10:30pm on February 15th and 8:30pm on March 15th.

One of the best deep sky objects of Pyxis is the one that also rises the earliest and is luckily at the highest declination. This is NGC2613, a moderately bright edge-on spiral galaxy. It is located at RA 8h 33m 22s Dec -22° 58' 24", or 3.5 degrees north northwest of Eta (η) Pyxidis in the northwest section of the constellation. It glows at magnitude 11.2 and presents itself as a pronounced spike of light. It spans 7.3' in its long dimension, and is 1.8' thick, with sharp edges along its length.

An interestingly similar trio of stars and star clusters repeats itself in the southern half of Pyxis. Once you master finding the first cluster, you should have no problem finding the other two, since they are all the same distance from marker stars of similar brightness. Starting with the cluster farthest to the north, NGC2627, first find magnitude 4.9 Zeta (ζ) Pyxidis. Then move 39' southwest to this magnitude 8 cluster. It is fairly large, covering 11', and has several bright (magnitude 11) stars interspersed with others ranging downward to magnitude 13 and fainter. NGC2627 shows a noticeable elongation. For your setting circles, NGC2627 is located at RA 8h 37m 18s Dec -29° 57' 00".

Move 3.5 degrees south southeast to 3.7 magnitude Alpha (α) Pyxidis as a starting point for finding NGC2658. This star cluster is 32' north of this brightest star of Pyxis. The cluster is small and faint (magnitude 9), but richer in stars than NGC2627. (RA 8h 43m 24s Dec -32° 39' 00".)

Near the southern boundary of Pyxis is a third cluster, 2.3 degrees south southwest from where we left NGC2658. Once again, start with Beta (β) Pyxidis and move practically the same distance as we have before (this time 38') to the northwest, or dial up coordinates RA 8h 38m 30s Dec -34° 46' 00". This cluster, designated NGC2635, is the most challenging. It is very dim at magnitude 11.2, and covering only 3' of sky. Many small telescopes will fail to resolve the 13th magnitude stars, but those with large apertures and high magnification will show a subtle but pleasant cluster in an irregular triangle shape.

Probably the most unique object in Pyxis is the one I've saved for last. Incidentally, if you're viewing Pyxis early in the evening, it will be the last object to come into view as Pyxis rises, since it is located in the far southeast corner of the constellation. NGC3818 is a combination star cluster and planetary nebula. Though it is not as striking as M46, also a cluster/planetary combination located nearby in Puppis, it is in the same category. This object sits in a very empty spot in the sky, so star hopping is difficult. If you can zoom in to coordinates RA 9h 16m 00s Dec -36° 37' 00", then do so. Otherwise, move about 7.5 degrees east of β Pyxidis. Look for a tight triangle of 4.6, 7, and 5.9 magnitude stars in your binoculars or finder scope, and then move 47' due north. About 9' of sky is covered by this delicate star cluster, and embedded within is a faint, 13th magnitude planetary nebula of about 40" in size. The nebula itself is not too difficult to spot since it has decent surface brightness, but a fairly large aperture is helpful. It is quite evident at high power in an 11" telescope with dark skies.

These cold nights of February present a wealth of stars and deep sky objects to view, even in the more withdrawn constellations like Pyxis. But braving the weather to meet the challenge of hunting down some of these dim objects makes for a rewarding night of observing.

Sagitta (sah jit' uh)
Abbreviation: Sge
The Arrow
Best observed between July 1st and September 30th

There have been several instances of arrows playing a part in mythical stories. Zeus's eagle was killed by an arrow, Apollo slew the Cyclops with an arrow, and Cupid, of course, uses arrows in his trade. All of these have been associated with Sagitta, a small summer constellation. Sagitta is approaching the meridian after midnight on early July nights near Altair, the most southern star of the Summer Triangle. Since it is immersed in the thick Milky Way, we don't see any galaxies. Instead, there are some interesting objects within our own galaxy.

Even though Sagitta is the third smallest constellation in the sky, it does have a Messier object, M71. This is an impressive globular cluster of magnitude 8.3 and a size of 7.2'. It's easy to find 21.5 arc minutes east of 6.2 magnitude 9 Sagittae, or half way between Gamma (γ) and Delta (δ) Sagittae (or RA 19h 53m 48s Dec +18° 47' 00"). You'll find a rich cluster of stars that range from magnitude 11 to 16. When M71 was first being studied, its true classification was in question. Some astronomers considered it a globular cluster, but others thought it to be a rich galactic cluster. When you observe M71 on your own, you'll notice that it does not have the dense central population of stars common among many globulars, but its distance of 18,000 light years and true diameter of 30 light years indicates that it probably is a loose globular.

About half a degree south southwest of M71 is a loose, scattered collection of stars classified as star cluster H20. It simply appears as a more concentrated portion of the Milky Way, with about 20 stars of magnitude 11, in a space about 10 arc minutes wide. H20 is near RA 19h 53m 10s Dec +18° 19'.

Out by itself, floating in the rich Milky Way, is planetary nebula IC4997. It is approximately 6 degrees southeast of Gamma (γ) Sagittae or RA 20h 20m 12s Dec +16° 45'. This is one of three dim planetary nebulae in Sagitta that are small and remote. This one glows at magnitude 12 and appears stellar except at very high magnification. The two other nebulae are NGC6886 (RA 20h 12m 42s Dec +19° 59'), and NGC6879 (RA 20h 10m 30s Dec +16° 55'.) All three of these are difficult to recognize due to their small size, but with enough magnification, they stand out from their high surface brightness.

An excellent Algol-type eclipsing variable star can be found in U Sagittae. You can find U Sagittae at RA 19h 18m 48s Dec +19° 36' 39", or 1.5 degrees west of 4 Vulpeculae, which is the brightest star in the hook of the asterism called "the coat hanger" in the neighboring constellation of Vulpecula. In the case of U Sagittae, we have a situation where a small, bright star is fully eclipsed by a larger but dimmer companion. Unlike many other variable stars, the behavior of U Sagittae is such that it can be observed in the course of a single evening if you catch it in or near its eclipse phase. Its full period is 80 hours, shining most of the time at magnitude 6.4. But it dims during a span of five hours to magnitude 9.1. Then it stays in eclipse for about two hours before quickly brightening again. A good comparison star of magnitude 8.2 is 11 arc minutes to the northeast.

Sagitta is small and dim, but should not be overlooked among the brighter areas of the summer Milky Way.

Sagittarius (saj it air' ee uss)
Abbreviation: Sgr
The Archer
Best observed between June 15th and August 15th

As the summer sky rises into view on balmy July and August evenings, we see the thickest part of our Milky Way galaxy forming a broad band stretching from roughly northeast to southwest. It is where this river of stars thickens to its greatest expanse in the southern regions, marking the direction of the center of our galaxy, that we find Sagittarius. The classical representation of Sagittarius, established by the Greeks and Sumerians, is that of an archer, typically a centaur (half man and half horse) with his bow and arrow aimed west toward Scorpius. The more modern image in the stars of Sagittarius (and also one that is easier to recognize) is of a teapot tilting to the southwest as if serving tea into an unseen cup, and the misty whiteness of the Milky Way rising northward is the "steam" emerging from the spout. Due to Sagittarius' privileged spot in the sky, it is packed with deep sky objects that can be viewed with a wide range of instruments, from small binoculars to the largest of backyard telescopes. Refer to the chart of Sagittarius near the front of this book for the best of the constellation's objects.

So, when going outside to observe Sagittarius, where does one begin? Perhaps approach it in a way that groups the constellation's three main categories of deep sky objects: globular clusters, open clusters, and emission nebulae.

At the bottom of the "teapot" are three easy to find globular clusters. M54 is 1.7 degrees west southwest of Zeta (ζ) Sagittarii. M70 lies midway between ζ and Epsilon (ε) Sagittarii, and M69 is 2.5 degrees directly west of M70. These three are similar in size and appearance, all between 7' and 10' in size, and near magnitude 8. Off the lid of the "teapot" is M28. This is a medium-sized globular, 1 degree northwest of Lambda (λ) Sagittarii, with a nicely defined core.

If you now slide just short of 3 degrees to the east northeast of M28, you'll see the incredible M22. This cluster is bigger and brighter than any other globular cluster on Messier's list. It is estimated to contain up to 1 million stars spanning about one hundred light years in diameter. Astronomers using the Hubble Space Telescope have seen evidence of wandering planet-sized objects as small as 80 times the mass of Earth swarming through the interior of M22. Another large cluster is M55, rivaling M22 in size and brightness. This one is to the southeast of the teapot's handle.

With so many of these clusters in close proximity in our sky, you can enjoy each one on an individual basis, or compare how one compares to the next. Here's a list of Sagittarius' globulars:

Object	RA	Dec	Size	Mag.
M54	18h 55.1m	-30° 28'	6'	8.7
M70	18h 43.2m	-32° 17'	4'	9
M69	18h 31.4m	-32° 21'	4'	8.9
M22	18h 36.4m	-23° 54'	22'	5.1
M28	18h 24.6m	-24° 52'	15'	8.5
M55	19h 40.0m	-30° 57'	15'	7.1

A variety of rich open clusters is found in the northern regions of Sagittarius. M24 gives you a real appreciation of how closely packed stars can appear when viewed from a distance. Your naked eye can see M24 as a glowing oval star cloud that is about 2 degrees long, and more than half a degree wide. The real treat is viewing it in a large aperture, wide field telescope. The tiny stars are like backlit spray from a garden mister and offer a hint at the staggering number of stars in a large galaxy such as ours. These points of light condense even further to form NGC6603. This tiny cluster packs about 100 12th magnitude stars in a round area within M24's boundaries. While you're here, make an attempt at Barnard 92, a dark nebula along the northwest edge of M24. The stars disappear behind this 12' x 6' blob of obscuring dust, thus revealing its presence.

M23 and M25 are two large clusters loosely dotted with stars. M23 is somewhat teardrop-shaped with over 100 stars of magnitude 7 and dimmer. M25 and M21 have about half that number of stars, with M25 covering about 4 times the amount of sky as M21. M18, one degree south of M17 (discussed below), is a very sparse cluster of about 15 stars of magnitude 8. Here are the open clusters of Sagittarius:

Object	RA	Dec	Size	Mag.
M23	17h 56.9m	-19° 01'	27'	5.9
M25	18h 31.7m	-19° 14'	37'	6.2
M24	18h 18.4m	-18° 25'	120' x 40'	4.5
M18	18h 19.9m	-17° 08'	7'	8.0
M21	18h 04.7m	-22° 30'	12'	7.2

Those fond of emission nebulae will not be disappointed here, either. M8, the Lagoon Nebula, is a huge cloud and associated star cluster showing a bright knot near the center, and a dark lane traversing its middle. NGC6530 is a cluster of young stars swimming within the nebulosity. M20, the Trifid Nebula, is a more challenging nebula but easily seen in dark skies. Its name comes from the dark lanes of dust trisecting the brightest part of the nebula. M17, the Swan or Omega nebula (near the borders of Scutum and Serpens Cauda) is a breathtaking cloud of hydrogen assuming the shape of a swan or check mark. Lastly, the archer's list of nebulae:

Object	RA	Dec	Size	Mag.
M8	18h 03.7m	-24° 23'	60' x 35'	5.0
M20	18h 02.4m	-23° 02'	29' x 27'	9.0
M17	18h 20.8m	-16° 10'	46' x 37'	6.9

By no means should you stop at the objects listed here. They are just the highlights of Sagittarius. A good star chart will keep you busy for hours with all this constellation has to offer.

Scorpius (skor' pee uss)
Abbreviation: Sco
The Scorpion
Best observed between June 1st and July 15th

It should be no surprise that this constellation is named after a scorpion. I can think of no other constellation that looks more like its namesake that this one. The bright red star, Antares, marks the scorpion's heart, and its tail curves down to the southeast to end in a stinger formed by the pair of stars, Shaula and Lesath. By the middle of July, Scorpius is crossing the meridian in the south sky as soon as the sky becomes fully dark, and is located west of Sagittarius, east of Libra, and below Ophiuchus. In mythology, Scorpius is the scorpion sent by the goddess Hera to attack Orion in order to disprove his invincibility. The scorpion's sting proved fatal, and we see this event played out every spring night when Scorpius is rising, and Orion is slowly falling on his side to set below the opposite horizon.

Due to its red color, Antares (Alpha (α) Scorpii) gets its name from the Greek components "anti" and "Ares", which when put together can be loosely translated to "rival" or "challenger" of Mars. When the two approach each other every time the orbit of Mars brings the planet to this part of the sky, the color of the two can be easily compared. In the family of stars, however, Antares has a different rival. It and Betelgeuse are the only two M-type supergiants of first magnitude brightness. If viewed at very high power, a sixth magnitude companion can be glimpsed through the glare of Antares 3 arc seconds distant.

Because much of Scorpius is immersed in the summer Milky Way, the number of galactic deep sky objects it contains compares only with two adjacent constellations, Sagittarius and Ophiuchus. Near Antares (1.3 degrees to the west) is the fantastic globular, cluster M4. Shining brightly at magnitude 5.9, this object (spanning 26.3') can be seen easily through a finder scope or binoculars, and is highly resolvable through a telescope. Even though it's one of the Milky Way's smallest globular clusters, its proximity to us makes it one of the best globulars of the summer sky. If you move one degree to the northeast, you'll come across NGC6144. Here is another globular cluster. Although it suffers a bit from its proximity to the far more spectacular cluster just mentioned, it is a decent globular, if judged on its own merits. It is a resolvable cluster measuring 9.3', and shining at magnitude 9.1 (only half a degree from Antares).

Another in the list of remarkable globular clusters is situated not quite half way between Antares and Beta (β) Scorpii. Here we see M80, a very rich collection of stars packed in a space of 8.9', and glowing with a magnitude of 7.2 (RA 16h 17m Dec -22° 59.6').

Moving toward the tail of Scorpius is yet another globular cluster, M62. This one practically straddles the boundary separating Ophiuchus from Scorpius, and some debate could be raised about to which constellation it belongs. Since it is in the area, consider it one of Scorpius'. Although smaller than M4, you might say that this one is the best of the group. At magnitude 6.6, it presents quite a sight in a telescope. Its bright, dense core softens to a speckled glow to round out its 14.1' size. Find M62 4.7 degrees northeast of Epsilon (ε) Scorpii, or RA 17h 01.2m Dec -30° 7'.

Between the middle part of the scorpion's tail and the stinger is an interesting planetary nebula, called the Bug Nebula, or NGC6302. While it glows only at magnitude 13, it is fairly easy to see in medium-sized telescopes because it measures only 0.8', so its surface brightness is moderately high. Look 4 degrees west from Lambda (λ) Scorpii, or center your scope on RA 17h 13.7m Dec -37° 06'. This does not fit the typical planetary nebula appearance, which is normally oval or round in shape, with perhaps a pair of lobed structures extending out into space. What we are seeing with NGC6302 apparently is a former star that has its rotational axis perpendicular with our line of sight. It has a pronounced pinch in the middle with material fanning out to each side. We can assume that this pinch is caused by a plane of dust, planets, or perhaps asteroids that inhibit the expansion of gasses through this area.

To the naked eye, the pair of open clusters M6 and M7 show as fuzzy blobs 4 degrees northeast of λ Scorpii. M6 (nicknamed the Butterfly Cluster) is a collection of 80 stars of magnitude 7 and dimmer tracing the shape of a butterfly's wings. Take special notice of the two lines of dim stars that can be pictured as the insect's antennae. M7 is about 4 times larger than M6 and contain stars of similar brightness. Binoculars are all you need to savor this cluster, but a telescope will show you the subtleties of the colors of these stars.

Sculptor (skulp' tohr)
Abbreviation: Scl
The Sculptor
Best observed between September 15th and November 1st

The French astronomer, Nicolas-Louis de Lacaille, is given credit for establishing this constellation, but it contains only a scattering of dim stars. His original name was L'Atelier du Sculpteur, the sculptor's workshop. Currently, only the sculptor remains. When it is highest in the sky, the constellation's southern boundary is on the horizon if viewed from 52 degrees north latitude, but it's up to 25 degrees high from the southern United States and similar latitudes. It reaches the meridian close to midnight during the middle days of September. You can find Sculptor east of Piscis Austrinus (marked by the bright 1st magnitude star Fomalhaut), and southwest of Cetus, the whale.

Although Sculptor does not set itself apart by having bright stars, it contains a wealth of very exciting deep sky objects. A good one to start with is NGC253. This edge-on galaxy sits high in the constellation near its border with Cetus. So impressive is this system that it is often referred to simply as the Sculptor Galaxy, and the 19th century Danish astronomer J. L. E. Dreyer effusively described it as, "very remarkable! Very, very bright, very, very large. . ." Indeed, when you point your telescope of whatever size at this bright galaxy, you'll see a magnitude 7.1 giant measuring 25' x 5'. This is nearly half a full moon in length. Large telescopes will have no difficulty seeing some of the clumps of dust littering its tight spiral arms and bright knots of dense stars. To find NGC253, you can start with magnitude 2 Diphda (Beta (β) Ceti, which is the southernmost bright star in the tail of Cetus), and move 7.5 degrees south and then 1.5 degrees east. Or you can dial up coordinates RA 00h 47.6m Dec -25° 17' 00". Either way, this galaxy should be readily apparent in either your finder scope or binoculars. There are few examples of a better, more easy to see edge-on galaxy than NGC253. It is also one of the nearest galaxies to us at about 9 million light years away.

A mere 1.7 degrees southeast of the Sculptor Galaxy is what could be called the Sculptor Globular Cluster because this cluster, NGC288, is very bright and impressive, as well. Measuring a wide 14' in diameter, and glowing at magnitude 8.1, this globular is also within the grasp of any backyard scope. Small instruments will show a soft glow, and large scopes will have no trouble resolving many of the stars of this loose cluster because they range from magnitude 12 to 16. Look to RA 00h 52.8m Dec -26° 35' 00' for this cluster. As a side note, the south galactic pole (SGP) is less than one degree southwest of this globular cluster.

On par with NGC253 is another edge-on galaxy, but it is very low in the sky. It makes its transit across the meridian exactly at 1am on September 1st, midnight on September 15th, and 11pm at the end of the month. These would be the best times to observe this galaxy, since it will be the highest in the sky, and away from the dust and atmospheric turbulence that will hinder the view. This galaxy is irregular in nature. It shows a mottled structure with uneven brightness across its length. Its western half is brighter than its eastern part and, in fact, is sometimes given two designations: NGC55 for the western clump, and IC1537 for the dimmer, eastern clump. Dimmer clouds of stars intervene between these two bright features. This is also a very nearby galaxy (a member of the Sculptor group), residing only about 8 million light years away.

NGC55 is 4 degrees south of magnitude 5.2 Theta (θ) Sculptoris, or RA 00h 14.9m Dec -39° 11' 00".

Another galaxy well within small backyard scopes is NGC300, which sits at RA 00h 54.9m Dec -37° 41' 00". This one is a face-on spiral, meaning that its magnitude 9.0 light is spread out across a large area of about 20' in diameter, making its surface brightness low. Most telescopes will show only its brighter nucleus, but large scopes will begin to bring out some of the brighter knots of stars in its arms. NGC300 is 1.7 degrees northwest of Zeta (ζ) Sculptoris.

NGC134 is a worthy target, due to its being easy to find. It sits about half a degree east southeast of a magnitude 4.6 star, Eta (η) Sculptoris. It's a rather bright magnitude 11.1 galaxy of a 8.4' x 2.0' size. This is another galaxy nearly edge-on to our view, also with mottled features in its spiral arms. NGC134 is situated at coordinates RA 00h 30.3m Dec -33° 14.7s.

Another in the long list of galaxies in Sculptor is NGC7793, a magnitude 9.6 face-on spiral with many tight spiral arms. It's a decent size at 9.4' x 6.3' located 3.7 degrees east of magnitude 5.3 Mu (μ) Sculptoris, or RA 23h 57.8m Dec -32° 35.4'.

An intriguing object, though beyond the grasp of most equipment other than long exposure images through large telescopes, lies halfway between Sigma (σ) Sculptoris and NGC300. It is identified as the Sculptor system, and is an extremely faint, sparsely populated dwarf elliptical galaxy that is a member of the local group. There are several other examples of this type of galaxy, such as the Fornax system and other systems in Leo, Draco, and Ursa Minor. All are nearby and would be invisible at much farther distances. This particular collection of stars covers a vast area of the sky, about 75', and is one of the intrinsically faintest galaxies in the heavens. It shines with a total luminosity of only 3 million suns. For a galaxy, that's an exceedingly small number. The earliest photographs of the Sculptor System showed such an elusive image of the galaxy that it was thought to be a photographic anomaly. Later, more modern images began to show that it was an actual object that was made of stars.

For a dim constellation, Sculptor contains many fantastic objects.

Scutum (skyoo' tum)
Abbreviation: Sct
The Shield
Best observed between June 15th and August 1st

Johannes Hevelius takes credit for naming this constellation. The year was 1690, and the Polish astronomer selected a group of stars lying north of Sagittarius and southwest of Aquila and named it "Scutum Sobiescianum," or Sobieski's shield. It was in honor of King John III Sobieski, who had recently won an important battle in the defense of Poland against the Turks, and Hevelius decided that the constellation should represent the king's coat of arms. Currently, it is simply known as Scutum. Look for Scutum to cross the meridian at 11pm July 15th about 35 to 50 degrees above the horizon (depending on your latitude) looking south.

The area of the sky in which Scutum resides is one of the more richly packed areas of the Milky Way. The Scutum star cloud, a particularly dense region, occupies the northern half of the constellation. The area benefits from a lack of obscuring dust as you look through the inner arms of the Milky Way Galaxy, allowing the light of the star cloud to come through relatively unimpeded. Here's where we find Scutum's best star cluster, M11, 2 degrees southeast of Beta (β) Scuti. This cluster lies in the foreground of the star cloud, and is one of the most densely populated clusters that you can hope to observe. Also known as the "Wild Duck" cluster after English observer William Smyth described it as "a flight of wild ducks," M11 shines at magnitude 5.8. The hundreds of individual stars within the cluster range in brightness from a single 8th magnitude star, slightly offset from the center, to approximately 16th magnitude. They are all packed into an area measuring 15'. M11 is approximately 3,000 light years away, and it has been around for 150 million years. The total intrinsic brightness of this magnificent cluster is equivalent to about 10,000 suns.

Looking south, this time 1 degree southeast of Epsilon (ε) Scuti (or RA 18h 45.2m Dec -9° 24'), we find M26. This is another galactic cluster containing about 30 stars also occupying an area of 15'. Its total magnitude is 8.0, and its brightest stars shine at magnitude 10.

A very easy star cluster to find is NGC6664. It's a mere one-half degree directly east of Alpha (α) Scuti, putting the two in the same field of view with low to moderate power. This one is a bit more scattered than M26, having about 40 stars that fill an area of 17' and assume a paisley shape. Even though this cluster is quite a bit more sparse than the others, it is away from the edge of the Scutum star cloud where the stars thin out, so the cluster's contrast with surrounding space enhances it some.

A star cluster of a different type is in the form of NGC6712. This globular cluster is 10 minutes in RA east and one-half degree south of fifth magnitude Epsilon (ε) Scuti (RA 18h 53.1m Dec -8° 42.0'). It's as bright as many of the globulars in Messier's catalog, so it's easy to pick out in small telescope, and you may even see it in your finder scope. It glows at magnitude 8.2. Even very small instruments can partially resolve this cluster, and large telescopes will present detail in the core.

Those of you with dark skies and large apertures may want to hunt for a faint planetary nebula, IC1295. It's a slightly elongated circle that is 1.5' across, glowing feebly at magnitude 15. If you go back to our globular cluster, NGC6712, slide your telescope 23' to the east southeast to RA 18h 54.6m Dec -8° 50'. At the center is a dim 17th magnitude stellar core which gave birth to the nebula.

R Scuti provides an interesting study of variable stars. Officially, it is classified as an RV Tauri type, characterized by more than one period overlapping another. An estimation of its period is 140 days, and most often its variations cover about one magnitude, from 5.0 to about 6.1. Every 4 to 6 oscillations, however, the minimum dips to magnitude 8. There are several adjacent stars that will provide good magnitude comparisons. Regular observations of this star are necessary if you want to catch it in one of these deep minima, since it begins to rise in brightness after only a few days. It's easy to find R Scuti. Look almost due south from Beta (β) Scuti just short of one full degree, or WNW from M11 just over one full degree to RA 18h 47.5m Dec -5° 42.3'. R Scuti will be about 1/3 of a degree southwest from the middle of a line connecting M11 and β.

Serpens Caput/Serpens Cauda (ser' pinz kah' put/kow' duh)
Abbreviation: Ser
Head/Tail of the Snake
Best observed between May 1st and July 15th

This constellation is uniquely broken into two parts, with Ophiuchus, representing the legendary physician Asclepius who learned how to bring the dead back to life, standing between them. To the west is Serpens Caput, the head of the snake, and to the east is Serpens Cauda, the tail of the snake. This complex pair of constellations lies just above the ecliptic, from the summer Milky Way to a distance west covering about 4 hours of RA. On July 15th, Serpens Caput begins to cross the meridian at about 8:30pm, and Serpens Cauda does the same at about 11:00pm. The head of the serpent looks toward Corona Borealis, while its tail reaches toward Aquila.

Starting at the head of the snake, we find M5, a spectacular globular cluster of magnitude 5.8 measuring 17.4' across. It is easy to find 22' NNW of the 5th magnitude star 5 Serpentis. (If you pause to notice 5 Serpentis, you'll see a double star of magnitudes 5 and 10 separated by 11".) M5 ranks among the best globulars in the sky, showing a bright glow in small scopes and a dazzling swarm of pinpoint stars in large instruments. It has a bright core that appears slightly oval in shape, and, along one edge, loops of stars that gives the impression of flower petals.

Up near the snake's head is NGC5962, a dim 12.2 magnitude galaxy, but one of the brightest among approximately a dozen galaxies in this half of Serpens. It can be found at RA 15h 36.5m Dec +16° 36.5', or not quite 4 degrees practically straight north of 3.8 magnitude Delta (δ) Serpentis. You should see a distinct nucleus surrounded by a gradually dimming fat oval shape completing the structure of the galaxy, about 3' across.

Slightly south of the line connecting Beta (β) and Gamma (γ) Serpentis is the long period variable star identified as R Serpentis. This star is a Mira class variable, named after Omicron (ο) Ceti (called Mira), the most notable variable star of this type in the constellation of Cetus. This star ranges in magnitude from about 6.9 to 13.4 during a period of about 356 days.

Moving over to Serpens Cauda, we see more galactic deep sky objects owing to the fact that here we are looking toward the Milky Way's spiral arms. M16 (RA 18h 18.8m Dec -13° 47.0') is very familiar to everyone who has looked through an astronomy publication. It's the photogenic Eagle Nebula, cataloged by Charles Messier because of the star cluster, but also designated NGC6611 from the nebulosity in which the stars are immersed. Any telescope will see the cluster, but those in the 10-inch range should easily bring out the nebulosity. Farther up, 4.5 degrees ENE of Theta (θ) Serpentis, is the open cluster IC4756. This is a large, loose cluster, magnitude 5 and 52' in size. An interesting feature of this cluster is the letters "mc" (or "wc" depending on its orientation in your optics) outlined in a few of the stars in the center of the cluster.

Now, move back to θ (named Alya), the star that marks the end of the snake's tail. This is a pleasant double star shining with a combined magnitude of 4.05. The two components share a common brightness and spectral type of A5, with an easy separation of 22".

Sextans (secks' tans)
Abbreviation: Sex
The Sextant
Best observed between March 1st and April 15th

This is an obscure little constellation that owes its creation to Johannes Hevelius, a 17th century German-Polish astronomer who is responsible for also naming other small, dim constellations (namely Canes Venatici, Lacerta, Leo Minor, Lynx, Scutum, and Vulpecula) to fill in some rather blank areas of the sky. He named this one after his astronomical sextant, a tool that allowed him to make a detailed star atlas which was published by his wife as a posthumous work three years after Hevelius's death in 1687.

Sextans lies south of the front half of Leo, and since it contains only 4th and 5th magnitude stars, a dark sky is necessary to adequately pick out its shape. It begins to rise at 7pm and reaches the meridian just after 10pm during the month of March. The handle of the familiar sickle shape that forms Leo's head and shoulders points straight to Alpha Sextantis, the constellation's brightest star.

Having a careful aim is important to finding the noteworthy objects in this area of the sky since there are no easily visible landmarks nearby to help guide the way. Star hopping from Regulus to the north or Alpha (α) Hydrae to the southwest will require sizable jaunts, but with perseverance, you will find some rewarding sights.

Let's start with 35 Sextantis, a double star whose components are of spectral type K3 and K0 and shine at magnitudes 6.5 and 7.5. Starting from 1st magnitude Regulus, move 6.5 degrees to the southeast to 4th magnitude Rho (ρ) Leonis, then another 2.5 degrees SSE to 5th magnitude 48 Leonis, and finally another 3 degrees to the SE to 35 Sextantis (or on your setting circles, RA 10h 43.3m Dec 4° 44.8'). Otherwise, envision 35 Sextantis forming a flat triangle with Regulus and ρ Leonis. This pair of yellowish-white stars shows a distinctive brightness contrast, and are separated comfortably by 6.5 seconds of arc. They form one corner of a trapezoidal pattern of four stars, having a shape almost identical to that of the trapezium at the center of the Orion nebula, only on a scale that's about 20 times larger. 35 will be the brightest of the four.

A small (but bright) galaxy sits in the southern environs of Sextans. It is just north of a line drawn from Epsilon (ε) to Gamma (γ) Sextantis, and about half-way between the two. Cataloged as NGC3115 but popularly named the "Spindle Galaxy," this is a lens shaped assemblage of stars measuring 4' x 1' and glowing at magnitude 10. Its high surface brightness is evenly spread across its face and allows it to withstand high magnification very well. It appears as a spiral seen edge on, but there is no appearance of dust lanes or any other features of differing contrast. Gently tapering on each side from its central mass are thin extensions that barely hint at a flat disk. The Spindle is located at RA 10h 5.2m Dec -7° 43.1'.

A pair of more challenging galaxies lies farther north. When you find them, you will see both sitting side by side less than 8 arc minutes apart. This pair of 11.5 magnitude galaxies are NGC3169 and NGC3166 and appear quite similar to each other. Both have a noticeable nucleus surrounded by very faint disks, covering only 4 to 5 arc minutes in their longest dimension. To track them down, start at α Sextantis and move 4 degrees north, and then six minutes of RA east. Or you may center your view at RA 10h 14m Dec +3° 26'.

Taurus (tah' russ)
Abbreviation: Tau
The Bull
Best observed between November 15th and January 30th

Winter nights bring some of the brightest stars the northern celestial hemisphere has to offer. During the early evening on any date in January, Taurus will have reached the meridian, and its brightest beacon, red Aldebaran (Alpha (α) Tauri), shines as the eye of the bull while the animal charges across the heavens toward Orion and Auriga. Since ancient times, Taurus has been seen as a bull, symbolizing strength and fertility. These qualities were attached to Taurus because at one time the constellation embraced the sun during the spring. This was important to early civilizations, since springtime marked the new agricultural period. Aldebaran serves as a marker for the Hyades star cluster, a large, naked-eye group of stars covering a patch of sky about 5 degrees across and forming a V-shaped pattern. However, Aldebaran is not a member of the group. It is merely a foreground star that happens to sit in the same line of sight as the cluster members. The Hyades is one of the closest clusters to Earth at a distance of about 130 light years. Some of the other stars scattered across the constellation are actually members of the Hyades if one takes into account their common proper motion through space, and as a group they are called the "Taurus Moving Cluster." The cluster has traveled beyond its closest point to our solar system, and several million years from now as it recedes in the distance, it will be seen as a small, run-of-the-mill cluster that will measure less than half a degree across located east of Betelgeuse in the constellation of Orion.

Another conspicuous and eye-catching cluster is M45, called the Pleiades, or "seven sisters." Its naked-eye appearance in the northwest part of the constellation is initially a fuzzy spot, but upon direct inspection, it breaks into six or seven (or more) distinct stars. The number depends on the visual acuity of the observer and atmospheric conditions. So striking is this cluster that it has its own mythological stories. One describes Zeus changing the sisters into celestial doves to escape Orion. Another says that the Pleiades was once the single brightest star in the heavens, and it became so boastful that the god Tane flung Aldebaran at it, breaking it up into the fragments that we see today. All of the brighter stars of the cluster are of spectral type B. This means they are hot and burning ferociously. With superbly dark skies and a large telescope, a wispy nebulosity can be seen surrounding some of the brightest members of the Pleiades, especially Merope. There has recently been some debate on whether the stars themselves are immersed in a dusty envelope, or whether the nebula is a foreground feature with the starlight shining through it during its journey to Earth. The stars of the Pleiades have individual names. Atlas and Pleione are the father and mother, and the seven sisters (Alcyone, Maia, Merope, Asterope, Taygete, Celaeno, and Electra) complete the family as labeled on the chart included at the front of this book.

The placid glow of the Pleiades contrasts with the violent beginnings of a nebula near the eastern edge of the constellation. On July 4, 1054, a new star blazed in the sky near Zeta (ζ) Tauri. Its appearance was recorded by Chinese astronomers as well as in Native American drawings. It slowly faded over the course of several months, and now we see in its spot M1, the Crab Nebula. It is located 1 degree northwest of ζ Tauri at RA 5h 34.5m Dec +22° 01'. The star that destroyed itself in the supernova explosion seen almost a thousand years ago is now a type of neutron star, called a pulsar, that is only a few miles in diameter and spinning 30 times per second. Although the pulsar is too dim to see, the nebula is an intriguing object to view. In small scopes, M1 appears as a ghostly oval smudge. Large instruments can capture the filaments of material which are slowly expanding into space. The magnitude 8.4 nebula spans 6' from a distance of 6,300 light years.

Objects of lesser fame are found in the eastern half of the constellation. Four star clusters are scattered here, and none are particularly rich. The two largest ones, NGC1746 and NGC1647, measure about 40' and contain between 25 and 50 stars collectively shining at magnitude 6 to 6.5. The two of these form an almost straight line from Aldebaran, with NGC1647 being 3.5° to the northeast of the bright star (or RA 4h 46.0m Dec +19° 04'), and NGC1746 sitting 9.8° away in the same direction (RA 5h 3.3m Dec +23° 49.0'). Farther south is another pair of clusters, NGC1817 and NGC1807. These are much smaller, each about 10'. NGC1807 holds about 15 stars of magnitudes 8 to 9. NGC1817, only 22' to the east-northeast of NGC1807, is significantly richer, containing about 50 stars of magnitudes 10 to 14. You can find these two clusters by also using Aldebaran as a starting point. Move 35 minutes in RA directly east (or 8.3°) from Aldebaran and you'll see the two clusters in the same low power field of view. NGC1807 is at RA 5h 10.7m Dec +16° 32.0', and NGC1817 resides at RA 5h 12.1m Dec +16° 43.0'.

Triangulum (try ang' yoo lum)
Abbreviation: Tri
The Triangle
Best observed between October 15th and January 15th

There are many fanciful stories from mythology that those from ancient civilizations tied to the stars to explain the origins of the constellations. Triangulum, by contrast, has a more worldly derivation. Just as we see a simple triangle in this constellation, so did the Romans who named this group of three stars Deltotum, because of its resemblance to the Greek letter Delta. Other explanations for why the stars of Triangulum caught the attention of early astronomers say that it was because the simple shape is reminiscent of the Nile delta, or the island of Sicily. But no matter what the origin, it has a fitting name. Triangulum is easy to spot near the zenith on November nights between the double arcs of stars forming Andromeda and the short hook pattern of Aries. It assumes the shape of a tall triangle.

If Triangulum was a rock band, then it would be a one hit wonder. That's because it has an object that outclasses every other point of interest within the constellation. This is M33, also known as the Pinwheel Galaxy. Along with the Andromeda Galaxy and our own Milky Way, M33 is the third large galaxy in the Local Group, which is composed of several galaxies of mostly dwarf sizes within 2 to 3 million light years of each other. M33, at 2.8 million light years distant, is slightly farther away than M31, and is also much smaller (about 50,000 light years across compared to over 100,000 for M31). And, adding to the contrast between the two, we see M33 more face on, while M31 is tilted more toward an edge-on orientation. What all this leads to is the fact that M33 covers a large part of the sky. It's about 60' x 40' in expanse, or well exceeding the area of the moon. This means that to capture the entire galaxy in an eyepiece, low power of 40 to 50 times magnification, is essential. In addition, its magnitude 6 light is spread over such a large extent that dark, transparent skies are also a requirement to best see its diffuse glow. It is this nature of M33 that makes it difficult to find at times.

With that said, point your telescope or binoculars a bit less than 4.5 degrees west of Alpha (α) Trianguli, the star that marks the apex of the triangle (or at RA 1h 33m 51s Dec +30° 39' 37"). M33 will show a soft glow at its nucleus that decreases gradually outward. It has two major arms that have a lumpy, convoluted structure and trace a stretched out "S" shape. With enough aperture and sky transparency, there is enough contrast to easily see its spiral form.

Once centered on M33, you can increase power to study some of the details within it. Its fat arms are knotted with star clusters and nebulae. One of the largest H II (ionized hydrogen) regions that is known in any galaxy is near the edge of M33's visible face. This is NGC604, an expansive hydrogen emission nebula and associated star cluster that is close to 1,000 light years in diameter. In medium-sized scopes, it will appear as a distinct fuzzy clump of material about 10' to the NE from the center of the galaxy. The chemistry of this region of M33 is similar to the Orion nebula. It should be easy to spot as the most prominent feature of the galaxy, besides the galactic nucleus itself.

It is easy to forget that Triangulum has other things to offer besides M33, but it's worth examining a couple of other objects before leaving this region of the sky. A nice, close double star, 6 Trianguli, is on the opposite side of the triangle from M33 below the line connecting Delta (δ) and Alpha (α) Trianguli. Use high power to overcome the 3.8" separation to see a magnitude 5.2 primary and 6.6 secondary. While some historical accounts of 6 Trianguli describe a yellow or gold primary and a blue or sapphire secondary, my experience with this double matches a small sampling of others' descriptions, and that is of two similar, pale yellow stars. The secondary shows a slightly gray-yellow tint. The spectra of these two stars are G5 and F6, which does indicate that some color difference may be apparent.

The best variable star in Triangulum is the long period variable R Triangulum. This star is four degrees east of the corner of the triangle marked by Gamma (γ) Trianguli (RA 2h 37m 16.6s Dec +34° 15' 54"). It is just at the limit of naked eye visibility at its maximum magnitude of 5.7, and then over a period of 266 days, it oscillates between that and a feeble magnitude 12.5.

Ursa Major (er' sah may' jer)
Abbreviation: UMa
The Great Bear
Best observed between March 1st and May 30th

Just about everyone is familiar with this constellation from the asterism of bright stars that outline the famous "Big Dipper". Show someone the night sky for the first time and often a question will be "Where's the Big Dipper?" Van Gogh has painted it, Shakespeare and Tennyson have mentioned it in their literature, and it should be safe to say that every civilization has taken note of it throughout history and prehistory. But the Big Dipper is only part of the complete constellation called Ursa Major. The most popular form that these stars have assumed in the eyes of observers is a bear. The Greeks called it "Arktos", the origin of our word "arctic". This is an appropriate derivation for a constellation that circles over the extreme northern latitudes. Greek legends explain that Callisto, who was a maiden that caught Zeus's eye, was turned into a bear by Zeus's jealous wife, Hera. Zeus gave the bear an honorary spot in the sky, but Hera had the last word by moving it near the pole so the bear would never enjoy rest, but endlessly circle the celestial pole. The two stars marking the outer side of the dipper's bowl can be used in finding Polaris, both the north star and also the star that marks the end of the handle of Ursa Minor (the Little Dipper). Follow these two stars starting at the bottom of the Big Dipper's Bowl and continue about the width of three fists at arm's length to Polaris.

Ursa Major is rich in objects to view. You can start with your naked eye by looking at the 2.3 magnitude star in the handle of the Big Dipper, where the handle bends (this is Zeta (ζ) Ursae Majoris, also called Mizar). Those with sharp eyes need no optical aid to spot a companion to Mizar of magnitude 4 about 12 arc minutes (or a little more than a third of the moon's diameter) away, called Alcor. Through a telescope, Mizar becomes a double itself, with a 4th magnitude companion 14" away. Not only was this the first double star to be discovered (in 1650), but it was also the first double to be photographed (1857). There is little or no color contrast between these stars. They all appear a pure white. Squeezing this trio of stars in your eyepiece makes for a truly striking sight. Another naked eye sight is found in three pairs of stars which, to the Arabs, represented the footprints of a leaping gazelle. As part of the bear, the stars form the toes of two of his hind feet and one front foot. Nu (ν) and Xi (ξ) form the first leap, Mu (μ) and Lambda (λ) the second leap, and Iota (ι) and Kappa (κ) the third.

Ursa Major lies far from the galactic equator, and this means that it is packed with galaxies. A pair of outstanding ones, M81 and M82, are in the northern part of the constellation centered near RA 9h 56m Dec

+69° 28'. Upon first sight, this duo becomes a favorite of many observers. They are separated by only 37 arc minutes, so both can be seen in a low power field. They have quite distinct personalities. M82 glows at magnitude 9.2 and has a size of 11.3' x 4.3'. Small telescopes will show a thin oval with perhaps a hint of dusty mottling toward the center. Large instruments will capture the inner calamities of this tortured galaxy. Whatever is happening in M82, it is a source of loud radio noise and is the cause of a tremendous explosion of material rushing out from its nucleus. This results in streamers and filaments similar to the those contributing to the appearance of M1, the supernova remnant in Taurus. M81, on the other hand, is a handsome spiral of magnitude 7.9. It is much larger than M82, covering 26' x 14'. Most will see a uniform oval, since only very large scopes will be capable of bringing out its spiral arms. An easy way to find this pair is to start at Gamma (γ) Ursae Majoris (the lower left star of the Dipper's bowl), move diagonally across the bowl to α, and then double that distance in a straight line to arrive at your target.

While we are in the neighborhood, you may want to move 46 arc minutes east to galaxy NGC3077, a decently bright galaxy (mag. 10.8) 5.3' x 4.4' in size. Being a dwarf elliptical, this galaxy shows a fat oval structure that is quite bright in the center, and then fades out to its boundaries.

Another pair of objects can be seen along the bottom surface of the Dipper's bowl. One and a half degrees ESE of Beta (β) Ursae Majoris is M108. This galaxy is very similar to M82 from the fact that it is oriented edge-on, measures 8.3' x 2.5, and has a dispersal of dusty lanes across its entire disk. M108 shines brightly at 10.1 and is located at RA 11h 11.5m Dec +55° 40'. Move only 48 arc minutes virtually in the same direction and you will see M97. A dark sky and at least a medium-sized telescope will reveal why this planetary nebula is given the popular name of the Owl Nebula. You will see two dark circles forming the eyes of the bird within the round face of the nebula with the 12th magnitude central star exactly in the middle. Unfortunately, the Owl glows at only magnitude 11.5 from a disk 3 arc minutes in diameter, so it can be elusive depending on seeing conditions.

Move toward Gamma (γ) Ursae Majoris to find another easy galaxy, number 109 in Messier's catalog (RA 11h 57.6m Dec +53° 23'). It is 38 arc minutes from the magnitude 2.4 star marking the lower left corner of the Dipper. This is a bright magnitude 9.8 barred spiral galaxy covering 7.6' x 4.9', so it is an easy target in any telescope. But 8 to 10 inches is necessary to attempt seeing details of the bar and spiral arms.

Forming a slightly flattened equilateral triangle with Zeta (ζ) and Eta (η) Ursae Majoris is the huge galaxy M101. It spans nearly half a degree across its face. Even though this galaxy shines strongly at magnitude 7.7, it has such a low surface brightness that little detail can be observed. I've spotted it in a telescope as small as 60 mm, but I had to look twice to make sure it was there. Larger scopes, of course, make it an easier target, but most views will still only show a soft glow with a slightly brighter middle. M101 sits at RA 14h 3.2m Dec +54° 21'.

Several other galaxies populate Ursa Major. Here's a partial list of some of the remaining brighter ones:

Object	RA/Dec	Mag.	Size
NGC2681	8h 53.5m/+51° 19'	10.9	3.6' x 3.4'
NGC2841	9h 22m/+50° 59'	10.1	8.1' x 3.5'
NGC2768	9h 11.5m/+60° 2'	10.8	8.1' x 4.3'
NGC3198	10h 19.8m/+45° 33'	10.8	8.6' x 3.3'
NGC3184	10h 18.3m/+41° 25.5'	10.4	7.4' x 7.0
NGC4605	12h 40m/+61° 36.5'	10.9	5.7' x 2.1

Let's end with an interesting double star, Xi (ξ) Ursae Majoris. The two stars average a comparatively small 26 AU of actual separation, but they are close enough to Earth that most telescopes can split them. Currently, they are separated by 1.8", but this will increase to a maximum of 3.1" by the year 2035. The total period is only 60 years, short enough to enable M. Savary in 1828 to be the first to compute the orbit of a binary star. Due to its relatively fast orbit, it is possible to see a distinct change in PA in a rather short time. If you have the patience to make prolonged periodic observations of these stars, you will be able to see their PA turn by about 90 degrees during a span of thirteen years.

So many more binary stars, variable stars, and galaxies that cannot be covered in a short description await you in Ursa Major. The only limitations are time and aperture.

Ursa Minor (er' sah my' ner)
Abbreviation: UMi
The Little Bear
Best observed between April 1st and August 1st

Without many bright stars, Ursa Minor might be just another overlooked constellation, but it holds a very important place in the sky and in legend. Greek folklore says that Ursa Minor is Arcas, son of the maiden, Callisto, who attracted the desires of Zeus. This aroused the jealousy of Zeus's wife, Hera, who turned Callisto and Arcas into bears.

Zeus then honored the bears by placing them in the sky. Callisto is Ursa Major and Arcas is Ursa Minor. This only angered Hera a second time, so she saw to it that the two bears circle the celestial pole endlessly without the luxury of setting below the horizon to seek rest. Ursa Minor is anchored to the celestial pole by the pole star, Polaris, which marks the tip of the little bear's tail or the end of the handle of the Little Dipper, the constellation's common name.

There have been a few different pole stars throughout history as the Earth's axis follows its 25,800 year long precession cycle. Currently that role is held by Polaris (Alpha (α) Ursae Minoris), but in centuries past, Beta (β) Ursae Minoris and Thuban (Alpha Draconis) have held that honor. In every case, the pole star has been seen with reverence as representing the anchor of the heavens, the summit of a cosmic mountain where the gods of the sky reside, a symbol of consistency and endurance, and in more modern times, a faithful guide star for ocean voyagers. A graphic is included elsewhere in this book that shows the stars near Polaris and illustrates the position of the north celestial pole (NCP), as well as the distance and direction of its movement over the next 60 years.

Detailed observations of Polaris show that it is both a double and a variable star. Its oscillations, however, are extremely small. They are on the order of 0.1 magnitude over a period of 4 days. Amateur observers will have more satisfaction viewing it as a binary star. The magnitude 2 (spectral type F8) primary star is separated by 18" from the magnitude 9 companion.

An interesting asterism can be found 1.8 degrees SSW from Epsilon (ε) Ursa Minoris near the 7.7 magnitude star, SAO2707. (For those of you with equatorial mounts, especially fork-mounted Schmidt-Cassegrains, you may want to turn the telescope 180 degrees so the telescope's axis is pointing to the celestial equator to make it easier to navigate so close to the pole.) It's called the "mini coat hanger" from its resemblance to the "big" coat hanger asterism in the Summer Milky Way running through the southwest part of the constellation of Vulpecula. While the larger one measures 1.5 degrees from end to end, this one measures a tiny 17'.

The best deep sky object Ursa Minor has to offer is the galaxy NGC6217. It's a spiral that measures approximately 2.5' x 2.0' and glows at magnitude 12.1. It lies about 2/3 of the way along a line from Epsilon (ε) to Eta (η), or RA 16h 32.6m Dec +78° 12'.

Virgo (ver' goh)
Abbreviation: Vir
The Maiden
Best observed between April 1st and June 1st

Throughout the millennia, Virgo has been seen as a female representing many different goddesses in various ancient cultures. Among them are Ishtar, the goddess of fertility in ancient Babylon; Isis, goddess of nature for the Egyptians; Astraea, Roman goddess of justice; the terrible Gorgon, Medusa; and an Anatolian goddess of the Earth called Cybele. Additionally, various Greek goddesses are represented by Virgo. These include Artemis, the Moon goddess; Athena, goddess of wisdom and war; Demeter, goddess of agriculture; and her daughter Persephone. The brightest star of Virgo, the hot blue star Spica of spectral type B1, is traditionally depicted as the stalk (or "spike") of wheat the maiden is holding in her left hand. We can find Virgo by following the method we use to also find Boötes. First, look north to the Big Dipper. Extend the curve of the dipper's handle to Arcturus, the brightest star in Boötes, and then continue traveling in the same direction to the brightest star in the southern part of the sky, in this case Spica in Virgo.

Virgo is synonymous with galaxies, but before we delve into these, let's start with NGC5634, a magnitude 9.6 globular cluster sitting midway and slightly below a line connecting Iota (ι) and Mu (μ) Virginis (RA 14h 29.6m Dec -5° 59'). This is a fairly bright globular visible even through light polluted skies, but it is rather small at 4.9' across. Through most amateur telescopes, it appears as a soft glow sitting within a triangle of dim stars.

We find the most impressive cluster of galaxies visible from Earth here in the constellation of Virgo. It contains more than 3,000 spiral and elliptical galaxies and is very close to us, cosmically speaking, sitting about 65 million light years away. Beginning our tour of galaxies, we can start with M104, a wonderful edge-on spiral popularly called the "Sombrero Galaxy." It is in a fairly empty part of the sky on the boundary between Virgo and Corvus. You can sweep 11.1 degrees west of Spica to find it, or center your scope on RA 12h 40m Dec -11° 37.3'. Its ghostly 8.3 magnitude glow is pierced by a well-defined dark lane around its large central bulge. Any telescope will provide a breathtaking view of this galaxy.

Moving north, we will see M61. Perhaps not quite as striking as M104, but still spectacular nonetheless. Start at Eta (η) Virginis, move north 4 degrees to the 5th magnitude star 16 Virginis, and then 1.2 degrees NNE to M61 (RA 12h 21.9m Dec +4° 28.3'). What you will find in M61 is a magnitude 10.1 face-on galaxy with a compact, almost stellar nucleus, with spiral arms forming a broad patch of light measuring 6.5' x 5.8'. Those of you will large apertures may see a curious feature about M61's arms. Instead of following graceful curves, they show sharp corners, each like the bend of an elbow, and produce a distorted diamond shape for the galaxy.

As we get closer to the core of the Virgo galaxy cluster, we see M49 4 degrees NNE of M61. Here is a bright (mag 9.1) elliptical system. Its center fades to its edges forming an oval measuring 10.3' x 8.4' at RA 12h 29.8m Dec +7° 59.96'.

The best sights in Virgo await us deep in the galaxy cluster which spills into the neighboring constellation of Coma Berenices to the north. There are actually some members of the cluster that extend as far north as Canes Venatici and south into Corvus. The most concentrated section of the cluster, containing the largest and brightest galaxies, is about 9 degrees west of Vindemiatrix, or Epsilon (ε) Virginis. An illustration near the front of this book shows the Messier galaxies that are scattered throughout the cluster. Many NGC galaxies fill in the spaces between these. For casual observing, however, it's not even necessary to know which galaxy is which (because that can be quite a challenge). Simply sweep your telescope through the cluster and admire galaxy after galaxy that come into view.

We can summarize Virgo's Messier galaxies as follows:

M86: This is a bright elliptical at RA 12h 26.2m Dec +12° 56.8', shining at magnitude 9.8.

M84: 17 arc minutes southwest of M86 at RA 12h 25.1m Dec +12° 53.3'. A similar elliptical of magnitude 10. NGC4388 is a magnitude 11.9 edge-on spiral to the south, forming an equilateral triangle with M86 and M84.

M87: Another elliptical, magnitude 9.6, at RA 12h 30.8m Dec +12° 23.4'. Two very small galaxies to the southwest can be seen in the same field.

M90: RA 12h 36.8m Dec +13° 9.8', magnitude 9.5. An elongated spiral with dusty mottling.

M89: RA 12h 35.7m Dec +12° 33', magnitude 9.8 elliptical.

M58: RA 12h 37.7m Dec +11° 49.2', magnitude 10.6. A spiral that's not quite turned face-on to us.

M59: RA 12h 42.1m Dec +11° 38.8', magnitude 10.6. An elliptical galaxy elongated north-south.

M60: RA 12h 43.7m Dec +11° 33'. Bright elliptical of magnitude 9.6. A 12th magnitude spiral, NGC4647, borders M60 3' to the northwest.

Vulpecula (vul peck' yoo luh)
Abbreviation: Vul
The Fox
Best observed between July 1st and September 30th

The last constellation alphabetically in our list of 64 constellations is the dim collection of stars which Johannes Hevelius originally named "Vulpecula cum Anser", or the "Fox with the Goose". In the roughly 300 years since he invented that name, the goose has been forgotten, and only the fox remains. Since it only contains one star with a Bayer designation (magnitude 4.4 Alpha (α) Vulpeculae, its brightest star), it is easily overlooked. To find this dim constellation, take advantage of one of summer's clear nights and look on the meridian around 1 am in July, or 8pm at the end of September. Between Albireo, the star marking the beak of Cygnus the swan, and the constellation Sagitta, the arrow (or the more conspicuous stars of Aquila), are the brightest stars of Vulpecula that shine between magnitudes 4.5 and 5. The rest of the constellation is a collection of even dimmer stars scattered to the east.

Vulpecula's claim to fame is that it is home to what many observers would call the most spectacular planetary nebula in the sky. This is M27, or the Dumbbell nebula. And, as the joke goes, it is NOT named after the astronomer who discovered it, but rather it gets its name from its unmistakable shape that contains two lobes similar to the weights of a dumbbell. Probably the easiest way to find M27 is to point your finder scope at the small, distinctive group of stars forming Sagitta. The two center stars, Zeta (ζ) and Delta (δ) Sagittae, fall in a line oriented toward the northeast. Follow this line in that direction about 4 1/2 degrees and your finder scope (or binoculars) will show you a small, ghostly disk of light. M27 is visible in a telescope of any size, but the degree to which you will recognize detail is determined by your aperture. Small scopes will pick out a roughly rectangular shape of gasses expanding away from an invisible central star. Telescopes in the 8-inch category will begin to see some mottling within the brightest areas of the Dumbbell and also give you a chance to spot the magnitude 13.9 central star. Larger instruments will additionally show you an unmistakable haze forming oval-shaped lobes extending in a perpendicular fashion to the bright dumbbell.

High power will fill your eyepiece with this very large nebula and give you a chance to spot some of the small detail within the knotty tangle of gasses. This material is being pushed away from the naked core of a former main sequence central star, an object that has a surface temperature close to 85,000 degrees Kelvin. The Dumbbell Nebula glows at magnitude: 7.6, has a size of 350", and is located at RA 19h 59.6m Dec

+22° 43'.

While M27 grabs your attention, most of Vulpecula's other deep sky objects, which are destined to be present since the constellation inhabits the dense summer Milky Way, are subtle by comparison. NGC6940, however, demands scrutiny as one of the better open clusters around. Immersed in the thickest part of Vulpecula's Milky Way, NGC6940 is a huge open cluster of more than a hundred dim stars 31' across, or as big as the full moon. Use low magnification on this cluster, otherwise, it will fill the view of your eyepiece and lose its identity. Find NGC6940 at RA 20h 34.6m Dec +28° 18'.

NGC6885 is another open cluster and slightly easier to find since it surrounds a magnitude 5.9 star, 20 Vulpeculae, that acts as the cluster's centerpiece. Here is a sprinkling of about 35 stars, ranging in brightness from 6 to 11, and arranging themselves in a shape resembling a spade or arrowhead around 20 Vulpeculae. The overall brightness of the cluster is magnitude 6 and is located at RA 20h 12.0m Dec +26° 29'.

Moving over to the southwest corner of the constellation, we find a large asterism whimsically nicknamed the "coat hanger", or more formally as Collinder 399. Oriented upside down, this collection of stars measures 1.5 degrees across and, indeed, looks like a simple coat hanger, complete with a conspicuous hook formed from four stars and six other stars forming the "shoulders" of the hanger.

If you follow the straight edge of the coat hanger toward the east, and you have at least a medium sized scope, you will see a star cluster with a truly different character than the other clusters described so far. It is difficult to see and would truly be a challenge to find if it weren't for the line of stars pointing directly at it. It lies 17' due east of the eastern most star of the coat hanger. This cluster, NGC6802, is formed from stars shining at magnitudes 13 through 18, and, therefore, very few (if any) stars can be resolved. The brightness of the entire cluster is magnitude 8.8, is bar-shaped, and measures 3' along its length. If you didn't know that you were looking at a galactic star cluster, you may mistake NGC6802 for a nearby irregular galaxy, as it will appear in most scopes only as a diffuse glow.

Appendix A

Locations of the planets, 2008-2025

The planets are as much a part of the night sky as the stars which make the fixed shapes and patterns that adorn it. They intrude through the constellations in a slow motion wander. If a point of light that wasn't there a few months or years ago appears in a constellation, or if you simply see a bright "star" that is not depicted on a star map, then it's most likely a planet. Below are the dates and locations (in terms of in which constellation the event occurs) of greatest eastern and western elongations (abbreviated GEE and GWE, respectively) of the inferior planets, and oppositions of the superior planets. This will guide you in identifying these wandering spots of light and knowing the best times to view them through the year 2025. Keep in mind that by no means are these dates the only time it's worth viewing the planets. They are still easily seen and worth viewing days and even weeks before and after these events. The phase transitions of Venus can make an interesting study of that planet during most of the time between conjunctions with the sun. Also, note the changes in brightness (that of Mars is the most obvious) of a planet as the distance between it and the Earth changes over time.

Through a telescope, Mercury and Venus will exhibit phases much like the moon as the geometry among these planets, the Earth, and the Sun changes with their orbital position. Mercury's appearance is a small disk and can suffer some from always being so close to the horizon. Venus is a brilliant white or off-white sphere and can become quite a large crescent in your eyepiece when it is near inferior conjunction with the sun (between the sun and Earth), or be small and gibbous near superior conjunction (on the opposite side of the sun from the Earth). For the best views, observe them closest to greatest eastern or western elongation when they will be farthest from the sun and, therefore, highest in the sky.

Visible features on Mars include the north and south polar ice caps and various light and dark markings across its surface. Jupiter's dynamic satellite system can be seen at low power, while high power will show various cloud features of the planet itself, including the great red spot, equatorial belts, and various other smaller weather systems in its turbulent atmosphere. The ring system of Saturn is easily seen at low magnification, as is also the ring's largest gap, the Cassini division, at higher power. The planet appears as a yellowish orb with subtle cloud bands and a coterie of several moons, with orange Titan being the brightest.

Uranus and Neptune are below naked eye visibility (to its credit, however, Uranus can sometimes be glimpsed without optical aid under dark skies) and appear as small, bluish or bluish-green orbs at high power. Since these two outer planets are not apparent to the naked eye, notable conjunctions with easily identified stars are listed with their opposition dates to aid in finding them. The distances of these conjunctions are listed for an observer along a latitude line of 33° north, but will not differ significantly when seen from other locations. The superior planets are best seen at opposition when their distances from the Earth are at their smallest and they are in the sky all night long, reaching their highest point in the sky at midnight.

Elongations of Inferior Planets

Mercury: Magnitude 0.4 to -0.6; distance from sun: 57.9 million kilometers; period of revolution around sun: 88 days; diameter: 4,878 kilometers; easily visible moons: none; apparent diameter: 5" to 13".
GEE Sept 11, 2008 (Virgo); GWE Oct 22, 2008 (Virgo)
GEE Jan 4, 2009 (Capricornus); GWE Feb 13, 2009 (Capricornus)
GEE April 26, 2009 (Taurus); GWE June 13, 2009 (Taurus)
GEE August 24, 2009 (Virgo); GWE Oct 6, 2009 (Virgo)
GEE Dec 18th, 2009 (Sagittarius); GWE Jan 27, 2010 (Sagittarius)
GEE April 8, 2010 (Aries); GWE May 26, 2010 (Aries)
GEE Aug 7, 2010 (Leo); GWE Sept 19, 2010 (Leo)
GEE Dec 1, 2010 (Sagittarius); GWE Jan 9, 2011 (Ophiuchus)
GEE March 23, 2011 (Pisces); GWE May 7, 2011 (Pisces)
GEE July 20, 2011 (Leo); GWE Sept 3, 2011 (Leo)
GEE Nov 14, 2011 (Scorpius); GWE Dec 23, 2011
GEE March 5, 2012 (Pisces); GWE April 18, 2012 (Pisces)
GEE July 1, 2012 (Cancer); GWE Aug 16, 2012 (Cancer)
GEE Oct 26, 2012 (Libra); GWE Dec 4, 2012 (Libra)
GEE Feb 16, 2013 (Aquarius); GWE March 31, 2013 (Aquarius)
GEE June 12, 2013 (Gemini); GWE July 30, 2013 (Gemini)
GEE Oct 9, 2013 (Libra); GWE Nov 16, 2013 (Virgo)
GEE Jan 31, 2014 (Aquarius); GWE March 14, 2014 (Capricornus)
GEE May 25, 2014 (Taurus); GWE July 12, 2014 (Orion)
GEE Sept 21, 2014 (Virgo); GWE Nov 1, 2014 (Virgo)
GEE Jan 14, 2015 (Capricornus); GWE Feb 24, 2015 (Capricornus)
GEE May 7, 2015 (Taurus); GWE June 24, 2015 (Taurus)
GEE Sept 4, 2015 (Virgo); GWE Oct 16, 2015 (Virgo)
GEE Dec 29, 2015 (Sagittarius); GWE Feb 7, 2016 (Sagittarius)
GEE April 18, 2016 (Aries); GWE June 5, 2016 (Aries)
GEE August 16, 2016 (Leo); GWE Sept 28, 2016 (Leo)
GEE Dec 11, 2016 (Sagittarius); GWE Jan 19, 2017 (Sagittarius)
GEE April 1, 2017 (Aries); GWE May 17, 2017 (Pisces)

GEE July 30, 2017 (Leo); GWE Sept 12, 2017 (Leo)
GEE Nov 24, 2017 (Ophiuchus); GWE Jan 1, 2018 (Ophiuchus)
GEE March 15, 2018 (Pisces); GWE April 29, 2018 (Cetus)
GEE July 12, 2018 (Cancer); GWE Aug 26, 2018 (Cancer)
GEE Nov 6, 2018 (Scorpius); GWE Dec 15, 2018 (Scorpius)
GEE Feb 27, 2019 (Pisces); GWE April 11, 2019 (Aquarius)
GEE June 23, 2019 (Gemini); GWE Aug 9, 2019 (Cancer)
GEE Oct 20, 2019 (Libra); GWE Nov 28, 2019 (Libra)
GEE Feb 10, 2020 (Aquarius); GWE March 24, 2020 (Aquarius)
GEE June 4, 2020 (Gemini); GWE July 22, 2020 (Gemini)
GEE Oct 1, 2020 (Virgo); GWE Nov 10, 2020 (Virgo)
GEE Jan 24, 2021 (Capricornus); GWE March 6, 2021 (Capricornus)
GEE May 17, 2021 (Taurus); GWE July 4, 2021 (Taurus)
GEE Sept 14, 2021 (Virgo); GWE Oct 25, 2021 (Virgo)
GEE Jan 7, 2022 (Capricornus); GWE Feb 16, 2022 (Capricornus)
GEE April 29, 2022 (Taurus); GWE June 16, 2022 (Taurus)
GEE Aug 27, 2022 (Virgo); GWE Oct 8, 2022 (Virgo)
GEE Dec 21, 2022 (Sagittarius); GWE Jan 30, 2023 (Sagittarius)
GEE April 11, 2023 (Aries); GWE May 29, 2023 (Aries)
GEE Aug 10, 2023 (Leo); GWE Sept 22, 2023 (Leo)
GEE Dec 4, 2023 (Sagittarius); GWE Jan 12, 2024 (Sagittarius)
GEE March 24, 2024 (Pisces); GWE May 9, 2024 (Pisces)
GEE July 22, 2024 (Leo); GWE Sept 5, 2024 (Leo)
GEE Nov 16, 2024 (Ophiuchus); GWE Dec 25, 2024 (Ophiuchus)
GEE March 8, 2025 (Pisces); GWE April 21, 2025 (Pisces)
GEE July 4, 2025 (Cancer); GWE Aug 19, 2025 (Cancer)
GEE Oct 29, 2025 (Scorpius); GWE Dec 7, 2025 (Libra)

Venus: Magnitude -3.7 to -4.4; distance from sun: 108.2 million kilometers; period of revolution around the sun: 225 days; diameter: 12,102 kilometers; easily visible moons: none; apparent diameter: 10" to 65".
GEE Jan 14, 2009 (Aquarius); GWE June 5, 2009 (Pisces)
GEE Aug 20, 2010 (Virgo); GWE Jan 9, 2011 (Libra)
GEE March 27, 2012 (Aries); GWE Aug 15, 2012 (Gemini)
GEE Nov 1, 2013 (Ophiuchus); GWE March 22, 2014 (Aquarius)
GEE June 6, 2015 (Cancer); GWE Oct 26, 2015 (Leo)
GEE Jan 12, 2017 (Aquarius); GWE June 3, 2017 (Pisces)
GEE Aug 17, 2018 (Virgo); GWE Jan 6, 2019 (Libra)
GEE March 24, 2020 (Aries); GWE Aug 13, 2020 (Gemini)
GEE Oct 29, 2021 (Ophiuchus); GWE March 20, 2022 (Capricornus)
GEE June 4, 2023 (Cancer); GWE Oct 23, 2023 (Leo)
GEE Jan 10, 2025 (Aquarius); GWE June 1, 2025 (Pisces)

Oppositions of Superior Planets

Mars: Magnitude 1.3 to -1.6; distance from sun: 227.9 million kilometers; period of revolution around the sun: 1.88 years; diameter: 6,786 kilometers; easily visible moons: none; apparent diameter: 4" to 26".

January 30, 2010 (Cancer)
March 5, 2012 (Leo)
April 9, 2014 (Virgo)
May 21, 2016 (Scorpius)
July 27, 2018 (Capricornus)
October 13, 2020 (Pisces)
December 7, 2022 (Taurus)
January 15, 2025 (Gemini)

Jupiter: Magnitude -1.8 to -2.6; distance from sun: 778.3 million kilometers; period of revolution around sun: 11.9 years; diameter: 142,984 kilometers; easily visible moons: Io, Europa, Ganymede, and Callisto; apparent diameter: 31" to 49".

August 13, 2009 (Capricornus)
September 21, 2010 (Pisces)
October 29, 2011 (Aries)
December 2, 2012 (Taurus)
January 5, 2014 (Gemini)
February 6, 2015 (Cancer)
March 8, 2016 (Leo)
April 8, 2017 (Virgo)
May 8, 2018 (Libra)
June 10, 2019 (Ophiuchus)
July 13, 2020 (Sagittarius)
August 20, 2021 (Capricornus)
September 26, 2022 (Pisces)
November 2, 2023 (Aries)
December 7, 2024 (Taurus)
January 9, 2026 (Gemini)

Saturn: Magnitude 0.6 to 0.0; distance from sun: 1.427 billion kilometers; period of revolution around sun: 29.46 years; diameter 120,536 kilometers; easily visible moons: Titan, Rhea, Iapetus, Dione, Tethys, Enceladus; apparent diameter: 15" to 21".

March 9, 2009 (Leo)
March 23, 2010 (Virgo)
April 4, 2011 (Virgo)
April 15, 2012 (Virgo)

April 28, 2013 (Libra)
May 11, 2014 (Libra)
May 23, 2015 (Libra)
June 2, 2016 (Ophiuchus)
June 14, 2017 (Ophiuchus)
June 27, 2018 (Sagittarius)
July 9, 2019 (Sagittarius)
July 20, 2020 (Sagittarius)
August 1, 2021 (Capricornus)
August 14, 2022 (Capricornus)
August 29, 2023 (Aquarius)
September 8, 2024 (Aquarius)
September 21, 2025 (Pisces)

Uranus: Magnitude 5.9 to 5.7; distance from sun: 2.87 billion kilometers; period of revolution around sun: 84 years; diameter: 51,118 kilometers; easily visible moons: none; apparent diameter: 3" to 4".
September 13, 2008 (Aquarius)
September 16, 2009 (Pisces)
50' north of Jupiter on September 18, 2010, late evening sky
September 21, 2010 (Pisces)
September 25, 2011 (Pisces)
September 29, 2012 (Pisces)
October 2, 2013 (Pisces)
October 7, 2014 (Pisces)
27' south of 5.2 magnitude Zeta Piscium on June 18, 2015, predawn sky
October 11, 2015 (Pisces)
October 15, 2016 (Pisces)
1.1 degrees north of 4.3 magnitude Omicron Piscium on June 18, 2017, predawn sky
October 19, 2017 (Pisces)
October 23, 2018 (Aries)
October 27, 2019 (Aries)
October 31, 2020 (Aries)
9' north of 5.8 magnitude Omicron Arietis on Oct 11, 2021, evening sky
November 4, 2021 (Aries)
November 8, 2022 (Aries)
November 14, 2023 (Aries)
November 16, 2024 (Taurus)
November 20, 2025 (Taurus)

Neptune: Magnitude 8.0 to 7.8; distance from sun: 4.5 billion kilometers; period of revolution around sun: 164.8 years; diameter: 49,528 kilometers; easily visible moons: none; apparent diameter: 2.5".

23' north of Jupiter and 35' NE of magnitude 5.0 Mu Capricorni on May 26, 2009

August 16, 2009 (Capricornus)
August 20, 2010 (Capricornus)
August 21, 2011 (Aquarius)
August 25, 2012 (Aquarius)

45' NW of 4.8 magnitude Sigma Aquarii on July 10, 2013, early morning

August 26, 2013 (Aquarius)
August 30, 2014 (Aquarius)
August 31, 2015 (Aquarius)
September 2, 2016 (Aquarius)
September 5, 2017 (Aquarius)

33' SSE of 3.7 magnitude Lambda Aquarii on Oct 16, 2017, evening sky

September 7, 2018 (Aquarius)

30" WSW of magnitude 4.2 Phi Aquarii on September 5, 2019, evening sky

September 10, 2019 (Aquarius)
September 10, 2020 (Aquarius)
September 15, 2021 (Aquarius)

September 16, 2022 (Aquarius)

3.7' NNW of magnitude 5.5 20 Piscium on September 11, 2023, evening sky

September 20, 2023 (Pisces)
September 20, 2024 (Pisces)
September 23, 2025 (Pisces)

Appendix B

Notable Star Names

Achernar (Alpha Eridani), magnitude 0.6
Acubens (Alpha Cancri), magnitude 4.3
Adhafera (Zeta Leonis), magnitude 3.4
Adhara (Epsilon Canis Majoris), magnitude 1.6
Albali (Epsilon Aquarii), magnitude 3.8
Albireo (Beta Cygni), magnitude 3.2
Alchiba (Alpha Corvi), magnitude 4.0
Alcor (visual double with Mizar, or Zeta Ursae Majoris), magnitude 4.0
Aldebaran (Alpha Tauri), magnitude 1.1
Alderamin (Alpha Cephei), magnitude 2.6
Alfirk (Beta Cephei), magnitude 3.2
Algedi (Alpha2 Capricorni), magnitude 3.6
Algeiba (Gamma Leonis), magnitude 2.6
Algenib (Gamma Pegasi), magnitude 2.9
Algol (Beta Persei), magnitude 2.8 (variable)
Alhena (Gamma Geminorum), magnitude 1.9
Alioth (Epsilon Ursae Majoris), magnitude 1.7
Alkaid (Eta Ursae Majoris), magnitude 1.9
Alkes (Alpha Crateris), magnitude 4.1
Almak (Gamma Andromedae), magnitude 2.2
Alnair (Alpha Gruis), magnitude 2.2
Alnilam (Epsilon Orionis), magnitude 1.7
Alnitak (Zeta Orionis), magnitude 2.0
Alphard (Alpha Hydrae), magnitude 2.2
Alphecca (Alpha Coronae Borealis), magnitude 2.3
Alpheratz (Alpha Andromedae), magnitude 2.1
Alrakis (Mu Draconis), magnitude 4.9
Alrisha (Alpha Piscium), magnitude 3.8
Alshain (Beta Aquilae), magnitude 3.9
Altair (Alpha Aquilae), magnitude 0.9
Altais (Delta Draconis), magnitude 3.1
Alterf (Lambda Leonis), magnitude 4.3
Aludra (Eta Canis Majoris), magnitude 2.5
Alula Borealis (Nu Ursae Majoris), magnitude 3.5
Alya (Theta1 Serpentis), magnitude 4.6
Ankaa (Alpha Phoenicis), magnitude 2.4
Antares (Alpha Scorpii), magnitude 1.2
Arcturus (Alpha Boötis), magnitude 0.2
Arneb (Alpha Leporis), magnitude 2.7
Ascella (Zeta Sagittarii), magnitude 2.6

Atik (Zeta Persei), magnitude 2.8
Azha (Eta Eridani), magnitude 3.9
Beid (Omicron1 Eridani), magnitude 4.0
Bellatrix (Gamma Orionis), magnitude 1.7
Betelgeuse (Alpha Orionis), magnitude 0.5 (variable)
Biham (Theta Pegasi), magnitude 3.5
Canopus (Alpha Carinae), magnitude -0.9
Capella (Alpha Aurigae), magnitude 0.2
Caph (Beta Cassiopeiae), magnitude 2.4
Castor (Alpha Geminorum), magnitude 1.6
Cebalrai (Beta Ophiuchi), magnitude 2.8
Chara (Beta Canum Venaticorum), magnitude 4.2
Chertan (Theta Leonis), magnitude 3.3
Cor Caroli (Alpha Canum Venaticorum), magnitude 2.9
Cursa (Beta Eridani), magnitude 2.8
Dabih (Beta1 Capricorni), magnitude 3.1
Deneb (Alpha Cygni), magnitude 1.3
Deneb Algedi (Delta Capricorni), magnitude 2.9
Denebola (Beta Leonis), magnitude 2.2
Dipda (Beta Ceti), magnitude 2.2
Dschubba (Delta Scorpii), magnitude 2.5
Dubhe (Alpha Ursae Majoris), magnitude 1.9
Edasich (Iota Draconis), magnitude 3.3
Elnath (Beta Tauri), magnitude 1.8
Eltanin (Gamma Draconis), magnitude 2.4
Enif (Epsilon Pegasi), magnitude 2.5
Errai (Gamma Cephei), magnitude 3.2
Fomalhaut (Alpha Piscis Austrini), magnitude 1.3
Furud (Zeta Canis Majoris), magnitude 3.0
Gemma (Alpha Coronae Borealis, also called Alphecca), magnitude 2.3
Giausar (Lambda Draconis), magnitude 3.8
Giena (Gamma Corvi), magnitude 2.8
Gomeisa (Beta Canis Minoris), magnitude 3.1
Grumium (Xi Draconis), magnitude 3.7
Hamal (Alpha Arietis), magnitude 2.2
Homam (Zeta Pegasi), magnitude 3.4
Izar (Epsilon Boötis), magnitude 2.4
Kaus Australis (Epsilon Sagittarii), magnitude 1.9
Kaus Borealis (Lambda Sagittarii), magnitude 2.8
Kaus Media (Delta Sagittarii), magnitude 2.7
Kitalpha (Alpha Equulei), magnitude 3.9
Kochab (Beta Ursae Minoris), magnitude 2.2
Kornephoros (Beta Herculis), magnitude 2.8
Kurhah (Xi Cephei), magnitude 4.3

Lesath (Upsilon Scorpii), magnitude 2.7
Marfik (Lambda Ophiuchi), magnitude 3.8
Markab (Alpha Pegasi), magnitude 2.6
Matar (Eta Pegasi), magnitude 2.9
Mebsuta (Epsilon Geminorum), magnitude 3.1
Megrez (Delta Ursae Majoris), magnitude 3.4
Meissa (Lambda Orionis), magnitude 3.4
Mekbuda (Zeta Geminorum), magnitude 4.0
Menkalinan (Beta Aurigae), magnitude 2.1
Menkar (Alpha Ceti), magnitude 2.8
Menkib (Xi Persei), magnitude 4.0
Merak (Beta Ursae Majoris), magnitude 2.4
Mesarthim (Gamma1 Arietis), magnitude 4.8
Mintaka (Zeta Orionis), magnitude 2.5
Mirach (Beta Andromedae), magnitude 2.4
Mirfak (Alpha Persei), magnitude 1.9
Mirzam (Beta Canis Majoris), magnitude 2.0
Mizar (Zeta Ursae Majoris), magnitude 2.4
Muphrid (Eta Boötis), magnitude 2.7
Muscida (Omicron Ursae Majoris), magnitude 3.4
Nashira (Gamma Capricorni), magnitude 3.7
Navi (Gamma Cassiopeiae), magnitude 2.2
Nekkar (Beta Boötis), magnitude 3.5
Nihal (Beta Leporis), magnitude 2.8
Nunki (Sigma Sagittarii), magnitude 2.1
Nusakan (Beta Coronae Borealis), magnitude 3.7
Phact (Alpha Columbae), magnitude 2.7
Phecda (Gamma Ursae Majoris), magnitude 2.5
Pherkad (Gamma Ursae Minoris), magnitude 3.0
Polaris (Alpha Canis Minoris), magnitude 2.1
Pollux (Beta Geminorum), magnitude 1.2
Procyon (Alpha Canis Minoris), magnitude 0.5
Propus (Eta Geminorum), magnitude 3.3
Rasalas (Mu Leonis), magnitude 3.9
Rasalgethi (Alpha1 Herculis), magnitude 3.5
Rasalhague (Alpha Ophiuchi), magnitude 2.1
Rastaban (Beta Draconis), magnitude 2.8
Regulus (Alpha Leonis), magnitude 1.3
Rigel (Beta Orionis), magnitude 0.3
Rotanev (Beta Delphini), magnitude 3.6
Ruckbah (Delta Cassiopeiae), magnitude 2.8
Rukbat (Alpha Sagittarii), magnitude 4.0
Sabik (Eta Ophiuchi), magnitude 2.6
Sadalmelik (Alpha Aquarii), magnitude 3.0

Sadir (Gamma Cygni), magnitude 2.3
Saiph (Kappa Orionis), magnitude 2.3
Scheat (Beta Pegasi), magnitude 2.6
Schedar (Alpha Cassiopeiae), magnitude 2.3
Seginus (Gamma Boötis), magnitude 3.0
Shaula (Lambda Scorpii), magnitude 1.7
Sheliak (Beta Lyrae), magnitude 3.5
Sheraton (Beta Arietis), magnitude 2.7
Sirius (Alpha Canis Majoris), magnitude -1.6
Skat (Delta Aquarii), magnitude 3.3
Spica (Alpha Virginis), magnitude 1.2
Sualocin (Alpha Delphini), magnitude 3.8
Sulafat (Gamma Lyrae), magnitude 3.3
Syrma (Iota Virginis), magnitude 4.0
Talitha (Iota Ursae Majoris), magnitude 3.1
Tania Australis (Mu Ursae Majoris), magnitude 3.1
Tania Borealis (Lambda Ursae Majoris), magnitude 3.5
Tarazed (Gamma Aquilae), magnitude 2.8
Thuban (Alpha Draconis), magnitude 3.6
Unukalhai (Alpha Serpentis), magnitude 2.6
Vega (Alpha Lyrae), magnitude 0.1
Vindemiatrix (Epsilon Virginis), magnitude 2.9
Wasat (Delta Geminorum), magnitude 3.5
Wasz (Beta Columbae), magnitude 3.1
Wezen (Delta Canis Majoris), magnitude 2.0
Yed Posterior (Epsilon Ophiuchi), magnitude 3.2
Yed Prior (Delta Ophiuchi), magnitude 2.7
Zaniah (Eta Virginis), magnitude 3.9
Zaurak (Gamma Eridani), magnitude 3.0
Zavijava (Beta Virginis), magnitude 3.6
Zosma (Delta Leonis), magnitude 2.6
Zubenelgenubi (Alpha Librae), magnitude 2.8
Zubeneschamali (Beta Librae), magnitude 2.7

Appendix C

Southern Hemisphere Constellations

Antlia (ant' lee uh), the air pump. Named by the 18th century French astronomer Nicolas-Louis de Lacaille after the compressed air pump that was invented by the British scientist Robert Boyle.

Apus (ay' puss), the bird of paradise. First established on Johann Bayer's star atlas in 1603.

Ara (Ah' ruh), the altar. Originally named Ara Centauri and considered to be the altar of the centaur Chiron.

Caelum (see' lum), the chisel. Named by Nicolas-Louis de Lacaille.

Carina (kuh ry' nuh), the keel. Originally part of the large constellation of Argo Navis, which represented the ship Argo that carried Jason and his Argonauts to search for the Golden Fleece, but now is broken off into its own constellation of just the ship's keel.

Chamaeleon (kuh mee' lee un), the chameleon. Created by Johann Bayer in 1603.

Circinus (sir' sin us), the drawing compass. Named by Nicolas-Louis de Lacaille.

Crux (krucks), the southern cross. A very bright pattern of stars seen as a cross by all southern explorers.

Dorado (doh rah' doh), the goldfish or dolphinfish. First appeared in Johann Bayer's 1603 star atlas.

Horologium (hoh roh loh' jee um), the clock. Named by Nicolas-Louis de Lacaille.

Hydrus (high' druss), the male water snake. First appeared in Johann Bayer's star atlas.

Indus (in' duss), the Indian. Named by Johann Bayer.

Mensa (men' suh), the table mountain. Named by Nicolas-Louis de Lacaille after the mountain of the same name near his observatory outside of Cape Town, South Africa.

Musca (muss' kuh), the fly. Named by Nicolas-Louis de Lacaille after incarnations of constellations originally established by Johann Bayer and Edmond Halley.

Norma (nor' muh), the level. Named by Nicolas-Louis de Lacaille.

Octans (ok' tanz), the octant. Originated by Nicolas-Louis de Lacaille.

Pavo (pah' voh), the peacock. Established by Johann Bayer in his 1603 star atlas.

Pictor (pick' tor), the painter. Named by Nicolas-Louis de Lacaille.

Reticulum (reh tick' yoo lum), the net. Named by Nicolas-Louis de Lacaille.

Telescopium (tell eh sko' pee um), the telescope. Named by Nicolas-Louis de Lacaille.

Triangulum Australe (try ang' yoo lum os tray' lee), the southern triangle. A constellation adapted by Johann Bayer for his 1603 atlas from a constellation originally envisioned by Pieter Theodor, a Dutch voyager.

Tucana (too kan' uh), the toucan. Created by Johann Bayer for his 1603 atlas.

Vela (vee' luh), the sail. A constellation broken off from the larger constellation of Argo Navis.

Volans (voh' lanz), the flying fish. Created by Johann Bayer in his 1603 atlas.

Glossary

Absolute Magnitude: A measure of a star's intrinsic brightness given in a value of how bright a star would appear from a standard distance of 10 parsecs.

Absorption Nebula: A gaseous, interstellar cloud which gives off no light of its own, but instead becomes visible by blocking light from background stars.

Angular Separation: The distance measured between two astronomical objects in degrees (or fractions of a degree, called arc minutes and arc seconds).

Altitude: A measure expressed in degrees of how high above the horizon a astronomical object appears.

Apastron: The farthest distance an orbit takes an object from a star, as with a planet or stellar companion.

Aperture: The diameter of a telescope tube that determines how much light the telescope will collect.

Apparent Magnitude: A measure of a star's brightness as it appears from the Earth.

Arc Minute: The angular distance in the sky equal to 1/60th of a degree.

Arc Second: The distance in the sky equal to 1/60th of an arc minute, or 1/3600th of a degree.

Asterism: A shape formed by a group of stars.

Astronomical Unit (AU): The average distance between the Sun and the Earth, or roughly 149,700,000 kilometers, or 93,000,000 miles.

Azimuth: Any direction pointing along the horizon measured from 0° to 359°, starting at due north and moving eastward.

Barred Spiral Galaxy: A galaxy featuring a prominent lane of stars bisecting the galaxy's disk.

Binary Star: A single star system composed of two stars orbiting one another.

Brown Dwarf: A very small star with a mass insufficient to initiate the fusion reactions required for the star to shine.

Cassegrain: A reflector telescope design consisting of a primary mirror and a smaller magnifying secondary mirror parallel to each other at opposite ends of the telescope tube.

Catadioptric: A telescope type utilizing both lenses and mirrors.

CCD: Charge-coupled device. This is an alternative to film for astro-photography, and utilizes an electronic chip to record a digital image.

Celsius: A temperature scale establishing 0 as the freezing point of water and 100 as the boiling point. One Celsius degree equals 1.8 Fahrenheit degrees.

Cepheid Variable: A variable star having a period directly proportional to its absolute magnitude.

Conjunction: A close pairing of two or more astronomical objects appearing in the sky, such as when two planets fall along the same line of sight so that they exhibit a small separation between them.

Constellation: A configuration of stars forming a defined shape.

Dark Nebula: See Absorption Nebula.

Declination: The mapping convention of objects in the sky used to establish their location on a north-south axis. Declination is expressed in degrees, minutes, and seconds.

Deep Sky: A term used to describe astronomical objects located beyond our solar system, including nebulae, star clusters, and galaxies.

Degree: 1/360th of any circle dividing the celestial sphere into two equal halves.

Dobsonian: A term given to a simple reflector telescope mounted on an alt-az base. It is named after its designer, John Dobson.

Eastern Quadrature: The position of a superior planet east of the sun when the geometry of the sun, Earth, and planet form a 90 degree angle. During this event, the planet will be seen in the evening sky.

Eclipsing Variable: A binary star that shows variations in brightness due to the fact that each star passes in front of the other with regular intervals, and blocks some or all of the background star's light.

Ecliptic: An imaginary line traced across the sky matching the plane of the solar system, and along which the sun and other planets travel.

Elliptical Galaxy: A very large galaxy with a spherical or oval shape.

Emission Nebula: A gaseous, interstellar cloud glowing with its own light resulting from nearby or embedded stars that fluoresces ionized hydrogen.

Eruptive Variable: A star which varies in brightness from regular or irregular explosions.

Filter: A glass or plastic accessory covering the telescope aperture or eyepiece. This either selectively allows certain wavelengths to pass through and others to be absorbed, or diminishes the overall intensity of light in the case of solar and lunar filters.

Focal Length: The distance the path of light travels through a telescope tube or eyepiece before reaching focus, express in millimeters. The focal length of a telescope divided by the focal length of the eyepiece equals the resulting magnification.

Galactic Equator: An imaginary line traced across the sky that matches the plane of the disk of the Milky Way Galaxy.

Galaxy: Any large aggregation of stars and other matter on the order of hundreds to tens of thousands of light years across, containing billions or trillions of stars.

Giant: A type of star many times the diameter of the sun.

Gigaparsec: 1 billion parsecs. One parsec equals 3.26 light years.

Globular Cluster: A large, spherical collection of stars held together by gravity and orbiting around the Milky Way galaxy.

Greatest Eastern Elongation: The position of an inferior planet when it is farthest east from the sun when viewed in Earth's sky. During this event, the planet will be seen in the evening sky.

Greatest Western Elongation: The position of an inferior planet when it is farthest west from the sun when viewed in Earth's sky. During this event, the planet will be seen in the morning sky.

Inferior Conjunction: The closest approach in the sky of an inferior planet to the sun from Earth’s viewpoint when the planet is between the Earth and the sun.

Inferior Planet: Either of the two planets (Mercury and Venus) that orbit closer to the sun than the Earth.

Index Catalog: An extension of the New General Catalog of astronomical objects.

Irregular Galaxy: A galaxy that is too small to organize itself into a spiral or elliptical variety and thus possesses an irregular shape.

Irregular Variable: A star that exhibits changing light output that follows no predictable pattern.

Kelvin: A temperature scale with each degree's range equal to that of a Celsius degree, with 0° being absolute zero, and +273° degrees being the freezing point of water.

Kiloparsec: One thousand parsecs.

Light Year: A unit of distance equal to the distance that light travels in one year, or 9.4 trillion kilometers.

Long Period Variable (LPV): A type of variable star which undergoes large changes in light output over a time span of several months.

Luminosity: A measure of how much light a star produces.

Luminosity Type: A classification of stars being either giants, super-giants, subgiants, main sequence, or white dwarfs.

Luminous Giant: A luminosity type describing a very large, very bright star.

Main Sequence: The luminosity type of a star that is in the middle of its life span and is characterized by stable, fusion reactions in its core. The sun is considered to be a main sequence star.

Maksutov: A catadioptric telescope design featuring a thick, meniscus lens at the aperture end of the telescope tube. Varieties include Maksutov-Newtonian and Maksutov-Cassegrain.

Megaparsec: One million parsecs.

Meridian: An imaginary line running across the sky, going from due north to due south, and running through zenith.

Messier Catalog: A catalog of 110 astronomical objects compiled by the 17th century French astronomer, Charles Messier.

Milky Way: A term referring to either the galaxy in which the sun resides, or the band of star clouds seen across the entire night sky.

Multiple Star: A star system consisting of three or more stars in a gravitationally-bound system.

Mythology: The set of stories involving Greek and Roman gods which attempted to explain certain natural phenomena, including the placement of the stars.

Nadir: The point directly below one's feet.

Nebula: A cloud of gas and/or dust in interstellar space.

New General Catalog (NGC): A compilation of thousands of astronomical objects observed by William Herschel and J.L.E. Dreyer in the 19th century.

Newtonian: A reflector telescope design utilizing one main, parabolic mirror and one flat, diagonal mirror.

North Celestial Pole (NCP): The point in the northern sky around which the celestial sphere rotates.

North Galactic Pole (NGP): The place in the sky where the north end of the rotational axis of the Milky Way points.

Open Cluster: A collection of stars in close proximity that is located in the spiral arms of the Milky Way Galaxy.

Opposition: The positioning of two astronomical objects when they are the farthest apart in the sky, commonly describing a planet that is opposite the sun.

Parallax: The phenomenon that makes a distant object shift position more slowly than a nearby object when the observer is in motion.

Parsec: A unit of distance equaling 3.26 light years.

Periastron: The point where an object's orbit brings it closest to a star, such as a planet or stellar companion.

Planetary Nebula: A circular nebula resembling a planet which results from the expansion of gasses from a dying star.

Position Angle (PA): The relative position of two objects, such as two stars in a binary system, when measured through a 360 degree circle.

Principle Galaxy Catalog (PGC): A catalog solely of galaxies compiled in the 20th century.

Precession: The natural wobble of the Earth's axis that turns in a complete circle during an interval of approximately 26,000 years.

Proper Motion: The visible motion of stars as seen from Earth relative to fixed points in the sky.

Pulsating Variable: A star that changes in brightness due to regular expansion and contraction of the star's physical size.

Red Dwarf: A small, cool star which glows brighter toward the red end of the visible spectrum.

Red Giant: A star that, toward the end of its lifetime, has had its outer atmosphere expand to many times the star's original size, and exhibits a red color due to the comparatively cooler visible surface.

Reflection Nebula: A nebula that is only visible by reflected light.

Reflector: A general telescope design using mirrors to focus light into an image.

Refractor: A general telescope design using lenses to focus light into an image.

Right Ascension (RA): The mapping convention of objects in the sky used to establish their location on an east-west axis. Right Ascension is expressed in hours, minutes, and seconds.

RR Lyrae Variable: A star that exhibits changes in brightness over the course of a few hours. The first type of this kind of variable star to be discovered is RR Lyrae.

Schmidt-Cassegrain: A Cassegrain telescope design using a thin corrector plate at its front aperture to aid in the proper focus of light.

Schmidt-Newtonian: A catadioptric telescope of a classic Newtonian design, but with a Schmidt corrector plate at the front aperture.

Seeing: A condition of the sky which determines how steady astronomical objects appear through a telescope and how much scintillation, or "twinkle," stars exhibit to the naked eye. Turbulence and differences in temperature through different layers of the atmosphere affect this.

Semi-major Axis: The average separation that exists between two stars of a binary system.

Separation: See Angular Separation.

Setting Circles: Dials attached to a telescope mount that display the declination and right ascension to which the telescope is pointing. There are analog and digital varieties of setting circles.

Sidereal: Referring to the stars. One application of this reference is the sidereal rotation of a planet, of the period between successive risings of a fixed star. The Earth's sidereal rotation is more than three minutes shorter than its solar rotation due to its motion around the sun, which shifts the position of the sun in the sky from one day to the next.

Sky Transparency: The quality the atmosphere above you that allows light from astronomical objects to reach your eye. The amount of clouds, dust, and haze affect the sky's transparency.

Solar Mass: A unit of measurement used to describe the amount of mass contained within a star, nebula, star cluster, black hole, or galaxy. It is the amount of mass contained within our sun.

South Celestial Pole (SCP): The point in the southern sky around which the celestial sphere rotates.

South Galactic Pole (SGP): The place in the sky where the south end of the rotational axis of the Milky Way points.

Spectral Type: Stellar categories that are determined by the star's surface temperature.

Spiral Galaxy: A common type of galaxy that has a spiral appearance with two or more arms curving around the center.

Star: An object made mostly of hydrogen and emitting energy due to fusion reactions in its core.

Star Party: an organized gathering of amateur astronomers and their telescopes with the purpose of sharing their observations.

Subgiant: A large star smaller than the largest stars, but larger than our sun.

Supergiant: The largest kinds of stars that exist. Many supergiants are over 100 times the mass of the sun.

Superior Conjunction: The closest approach in the sky of an inferior planet to the sun in the Earth's sky when the planet is on the opposite side of the sun as the Earth.

Superior Planet: Any of the planets that orbit farther from the sun than Earth. Mars, Jupiter, Saturn, Uranus, and Neptune are superior planets.

Supernova: Either of two types of large stellar explosions resulting from the collapse of a large star's core (type II), or the sudden fusion of accumulated hydrogen drawn from a nearby companion star on a white dwarf's surface (type Ia).

Supernova Remnant: The visible gaseous debris remaining after a large star explodes as a supernova.

Twilight: The period of semi-darkness after sunset. Twilight is divided into three phases: Civil, Nautical, and Astronomical.

Universe: The term given to all that is known to exist within and beyond our Milky Way galaxy.

Variable Star: Any type of star exhibiting changes in brightness.

Western Quadrature: The position of a superior planet west of the sun when the geometry of the sun, Earth, and planet form a 90 degree angle. During this event, the planet will be seen in the morning sky.

White Dwarf: The remnant that remains after a star of similar mass to our sun reaches the end of its life and has shed its outer atmosphere to reveal its white-hot core.

Zenith: The point in the sky that is directly above one's head.

Zodiac: The collection of the twelve constellations that lies along the ecliptic.

Zodiacal Light: The cone-shaped band of diffuse light seen shortly after the sky becomes fully dark, or before the first light of dawn, that results from sunlight scattering off dust in the plane of our solar system. This is best seen during spring and fall when this dust plane is more perpendicular to the horizon.

References

Burnham, Robert Jr. *Burnham's Celestial Handbook: An Observer's Guide to the Universe Beyond the Solar System.* New York: Dover Publications, Inc., 1978.

Burnham, R., Dyer, A., Garfinkle, R. A., George, M., Kanipe, J., & Levy, D. H. *A Guide to Advanced Skywatching: The Backyard Astronomer's Guide to Starhopping and Exploring the Universe.* San Francisco: Fog City Press, 1997.

Chartrand, Mark R. *National Audubon Society Field Guide to the Night Sky.* New York: Chanticleer Press, 1997.

Edmund Mag 5 Star Atlas. Barrington, NJ: Edmund Scientific Co., 1974.

Tirion, Wil. *Sky Atlas 2000.* Cambridge, MA: Sky Publishing Corporation, 1981.

The Sky Astronomy Software. Golden, CO: Software Bisque, 1999.

ISBN 142514066-1

9 781425 140663

Made in the USA
Lexington, KY
30 November 2009